Methods for Cutaneous Investigation

COSMETIC SCIENCE AND TECHNOLOGY SERIES

Series Editor
ERIC JUNGERMANN
Jungermann Associates, Inc.
Phoenix, Arizona

Volume 1: Cosmetic and Drug Preservation: Principles and Practice, *edited by Jon J. Kabara*

Volume 2: The Cosmetic Industry: Scientific and Regulatory Foundations, *edited by Norman F. Estrin*

Volume 3: Cosmetic Product Testing: A Modern Psychophysical Approach, *by Howard R. Moskowitz*

Volume 4: Cosmetic Analysis: Selective Methods and Techniques, *edited by P. Boré*

Volume 5: Cosmetic Safety: A Primer for Cosmetic Scientists, *edited by James H. Whittam*

Volume 6: Oral Hygiene Products and Practice, *by Morton Pader*

Volume 7: Antiperspirants and Deodorants, *edited by Karl Laden and Carl B. Felger*

Volume 8: Clinical Safety and Efficacy Testing of Cosmetics, *edited by William C. Waggoner*

Volume 9: Methods for Cutaneous Investigation, *edited by Robert L. Rietschel and Thomas S. Spencer*

Volume 10: Sunscreens: Development, Evaluation, and Regulatory Aspects, *edited by Nicholas J. Lowe and Nadim A. Shaath*

Other Volumes in Preparation

Methods for Cutaneous Investigation

edited by

Robert L. Rietschel
Ochsner Clinic
New Orleans, Louisiana

Thomas S. Spencer
Cygnus Research Corporation
Redwood City, California

MARCEL DEKKER, INC.   New York and Basel

Library of Congress Cataloging-in-Publication Data

Methods for cutaneous investigation / edited by Robert L. Rietschel, Thomas S. Spencer.
    p. cm. -- (Cosmetic science and technology series ; v. 9)
  Includes bibliographical references.
  ISBN 0-8247-8264-X (alk. paper)
  1. Skin--Research--Methodology. 2. Dermatotoxicology--Methodology. 3. Drugs--Testing--Methodology. 4. Skin--Diseases--Animal models.   I. Rietschel, Robert L. II. Spencer, Thomas S., III. Series.
QP88.5.M47
616.5'027--dc20
                                                                90-2892
                                                                   CIP

This book is printed on acid-free paper.

Copyright © 1990 by MARCEL DEKKER, INC. All Rights Reserved

Neither this book nor any part may be reproduced or transmitted in any form or by any means, electronic or mechanical, including photocopying, microfilming, and recording, or by any information storage and retrieval system, without permission in writing from the publisher.

MARCEL DEKKER, INC.
270 Madison Avenue, New York, New York 10016

Current printing (last digit):
10 9 8 7 6 5 4 3 2 1

PRINTED IN THE UNITED STATES OF AMERICA

We dedicate this work to Colonel William A. Akers, U. S. Army Medical Corp, who assisted both of us in our early dermatological investigative careers. In the same spirit that Dr. Akers has helped ourselves and other young investigators begin a career of searching for new information and understanding to benefit the care of patients with skin disease, we dedicate this book to future young investigators with hope that it will smooth their pathway as Dr. Akers smoothed ours with words of encouragement and understanding.

Introduction to the Series

The Cosmetic Science and Technology series was conceived to permit discussion of a broad spectrum of current knowledge and theories of cosmetic science and technology. The series is made up of books written either by a single author or by a number of contributors. Well-known authorities from industry, academia and the government participate.

The aim is to cover the many facets of cosmetic science and technology. Topics will be drawn from a wide range of disciplines ranging from chemical, physical, analytical, and consumer evaluations to safety, efficacy, and regulatory questions. Organic, inorganic, physical, and polymer chemistry, emulsion technology, microbology, dermatology, toxicology, and so on, all play a role in cosmetic science. There is little commonality in the scientific methods, processes, or formulations required for the wide variety of cosmetics and toiletries manufactured. Products range from hair care, oral care, and skin care preparations to lipsticks, nail polishes and extenders, deodorants, body powders, aerosols to over-the-counter products, such as antiperspirants, dandruff treatments, antimicrobial soaps, acne products, and suntan lotions.

Cosmetics and toiletries represent a highly diversified field with many subsections of science and "art". Even in these days of high technology, a lot of "art" and instinct is used and needed in the formulation, evaluation, and selection of cosmetic products, although there is a strong move toward the "scientific method," particularly in such areas as claim substantiation, product evaluation, and analysis.

Emphasis is placed on reporting the current status of cosmetic technology and science in addition to historical reviews. The series will include books on safety, efficacy testing, oral hygiene products, sunscreens, deodorants and antipersperants, hair care, and claim substantiation. Contributions range from highly sophisticated and scientific treaties, to primers, practical applications, and pragmatic presentations. Authors were encouraged to present their own concepts as well as well-established theories. They have been asked not to shy away from fields that are still in a state of development or transition, nor to hesitate to present detailed discussions of their own work. Altogether, we intend to develop in this series a collection of critical surveys by noted experts covering most phases of the cosmetic business.

The tenth book in the series, *Methods for Cutaneous Investigation*, edited by Robert L. Rietschel and Thomas S. Spencer, comprises twelve chapters discussing some of the most important techniques for measuring and evaluating the properties of the skin. A variety of human, animal, and instrumental test methods are described by leading experts in the field. The topics are of particular importance at a time when the cosmetic industry is striving to utilize more scientific methods for evaluating and supporting the claims made for their products, which nowadays are often designed to provide both functional treatment as well as cosmetic effects.

I want to thank the contributors and editors who are participating in this series, the editorial staff at Marcel Dekker, Inc., and above all, my wife, Eva, without whose editorial help and constant support I would never have undertaken this project.

Eric Jungermann, Ph.D.

Preface

Each year, bright and energetic young minds enter a variety of disciplines that affect the field of dermatological research. Some are dermatology residents, that is, physicians in training who will, because of their various backgrounds, develop new ideas for the management of skin disease. They well may wish to test these ideas during their training. Others may be in the applied sciences looking to improve the appearance of skin and its ability to withstand the adverse environmental factors to which it is exposed and to test their own ideas on improved methods of skin management for either cosmetic or practical purposes.

    This book is an attempt to assist the bright young minds looking to test new products, ideas, and skin management programs so that they need not necessarily "reinvent the wheel" with their attempts to prove their points. It is hoped that they will find within the pages of this book guidelines on what types of methods have been found most effective and most reliable for monitoring the parameters that they are interested in studying. This book is designed to assist the younger investigator or an investigator switching to a new field, and it may not be as useful to the experienced investigator who is already familiar with the background literature in his or her chosen field of research. It should help facilitate access to the broader dermatological literature and more detailed reports on technique, with minimal amounts of wasted energy.

Robert L. Rietschel
Thomas S. Spencer

Contents

Introduction to the Series   *v*
Preface   *vii*
Contributors   *xi*

1. Human Test Panels   1
   John E. Wild and Marydale Murnan

2. Dermatotoxicology: Animal Irritancy Testing   19
   Susan S. Roper

3. "Predictive" Animal Sensitization Assays: What Do They Mean?   47
   Klaus E. Andersen and Howard I. Maibach

4. Predictive Animal Phototesting   85
   Robert L. Rietschel

5. Human Predictive Phototesting — 93
   Edward A. Emmett and Marcia A. Levy

6. Measurement of Cutaneous Blood Flow — 107
   Dirk B. Robertson and Howard I. Maibach

7. Skin Impedance Measurement — 121
   William I. Archer, R. Kohli, J. M. C. Roberts, and Thomas S. Spencer

8. Techniques for Sampling the Bacterial Flora of the Skin — 143
   Anthony A. Gaspari

9. Skin Permeation: In Vitro Techniques — 171
   Robert L. Bronaugh

10. The Evaluation of Acne in Humans — 185
    Robert L. Rietschel

11. Transepidermal Water Loss: Methods and Applications — 191
    Thomas S. Spencer

12. Assessing Hair Growth in Male Pattern Baldness — 219
    Robert L. Rietschel

Index — 225

Contributors

Klaus E. Andersen Department of Dermatology, University Hospital, Odense, Denmark

William I. Archer* Department of Physical and Analytical Chemistry, European Research and Development Center, Johnson Wax Ltd., Egham, England

Robert L. Bronaugh Division of Toxicological Studies, Food and Drug Administration, Washington, D.C.

Edward A. Emmett[†] Division of Occupational Medicine, The Johns Hopkins Medical Institutions, Baltimore, Maryland

---

Present affiliations:
*Department of International Consumer Products, Research and Development, S.C. Johnson & Son, Inc., Racine, Wisconsin.
†National Occupational Health and Safety Commission, Sydney, Australia.

Anthony A. Gaspari* Department of Dermatology, Emory University School of Medicine, Atlanta, Georgia

R. Kohli[†] European Research and Development Center, Johnson Wax Ltd., Egham, England

Marcia A. Levy Center for Occupational and Environmental Health, The Johns Hopkins University, Baltimore, Maryland

Howard I. Maibach Department of Dermatology, University of California Hospital, San Francisco, California

Marydale Murnan Department of Clinical Research, Hill Top Research, Inc., Cincinnati, Ohio

Robert L. Rietschel Department of Dermatology, Ochsner Clinic, New Orleans, Louisiana

J. M. C. Roberts European Research and Development Center, Johnson Wax Ltd., Egham, England

Dirk B. Robertson Department of Dermatology, Emory University School of Medicine, Atlanta, Georgia

Susan S. Roper[‡] Department of Dermatology, Emory University School of Medicine, Atlanta, Georgia

Thomas S. Spencer Department of Research and Development, Cygnus Research Corporation, Redwood City, California

John E. Wild Department of Clinical Research, Hill Top Research, Inc., Cincinnati, Ohio

---

Present affiliations:
*Department of Dermatology, Strong Memorial Hospital, University of Rochester, Rochester, New York.
[†]Colgate Palmolive, Herstal Milmort, Belgium.
[‡]Division of Dermatology, Department of Internal Medicine, University of South Florida School of Medicine, Tampa, Florida.

Methods for Cutaneous Investigation

1
Human Test Panels

JOHN E. WILD and MARYDALE MURNAN *Hill Top Research, Inc., Cincinnati, Ohio*

INTRODUCTION

The use of human subjects in research has occurred since the early days of medical practice. A historical review of the use of human subjects will give the reader a background for understanding the principles now in practice for employing human subjects in experimentation.

The use of human subjects in research has increased dramatically in the past quarter-century. This increase in the use of subjects is directly proportional to the promulgation of regulations requiring the determination of the safety and efficacy of prescription drugs, over-the-counter drugs, devices, cosmetics, and chemicals either already on the market or scheduled to enter the marketplace.

Our goal here is to give a brief historical background as to the use of human subjects, the ethical means for their proper use, and guidelines for carrying out successful human projects.

HISTORICAL PERSPECTIVE ON THE USE OF HUMAN SUBJECTS

The early Egyptians performed vivisection on criminals in order to learn more about body functions. Later, the Roman physician Celsus defended such experimentation in his statement "It is not cruel to inflict on a few criminals sufferings which may benefit multitudes of innocent people throughout all centuries" (Spencer, 1935–1938) (1, p. 4). The Egyptians as well as the Romans saw benefit to society in human experimentation; however, the consequences to the subject were not given any consideration.

Through the centuries not a great deal of well-documented human experimentation took place. In the late nineteenth century Walter Reed performed his work on yellow fever with subjects who had consented to be exposed to areas that were infested with mosquitos.

There were those who believed that Reed's work was unethical and immoral; however, others believed that since the subjects had given their consent and were aware of the circumstances, the research was ethical.

The above examples have established two premises as a requisite for ethical research, namely, benefit to society and voluntary consent of subjects.

Clinical experimentation began to take shape in the early twentieth century. "It was only during the 1920s that the model of 'investigator–clinician' was shaped. In the early 1930s the methodological contributions of Sir Bradford Hill and Sir Ronald Fisher provided essential statistical tools for the design and analysis of clinical experiments. By the late 1930s, the professional clinical investigator had become established on the medical scene, and research had become an integral part of hospital practice (Resider, 1978) (1, p. 4).

Human experimentation led to many medical advances in a short period of time. The introduction of antibiotics, antiseptic practice, immunization, and anesthesia came through human experimentation. These innovations were accepted by the community as great steps forward to benefit mankind. An era of acceptable clinical research had arrived.

Then in 1945, the news, both written and photographic, of the horror of the Nazi prison camps spread through the world. German physicians had committed crimes against human subjects that would change clinical investigation forever (1, p. 5).

Other abuses of experimental subjects have been revealed since the Nazi prison camps. These include experiments in the United

States at the Jewish Chronic Disease Hospital, where senile patients unknowingly were injected subcutaneously with cancer cells; at Willowbrook State Hospital, where retarded children were infected; and the Tuskegee Syphilis Study in which 300 black men were not treated for diagnosed syphilis even though effective antibiotic treatment was available (1, p. 5).

These, as well as many other similar human experimentation events, led to the introduction of codes of practice in order to protect human subjects. The initial attempts at human subject protection were in the Nuremberg Code of 1949. In 1964 the World Health Organization adopted guidelines for the protection of human subjects, in the Helsinki Declaration. In 1974, the United States through PL93-348 established the National Commission for the Protection of Human Subjects. Their report served as basis for the regulations published in the Federal Register in January 1981 titled *Protection of Human Subjects; Informed Consent; Standards for Institutional Review Boards for Clinical Investigations and Clinical Investigations Which May Be Reviewed by Expedited Review Procedure.*

The U.S. Department of Health and Human Services through the Food and Drug Administration continues to update the actual implementation of the regulations through inspection of clinical research projects and clarification of regulatory interpretation.

The basic principles for guiding international biomedical research involving human subjects are contained in the Declaration of Helsinki as adopted by the 18th World Medical Assembly in 1964 and revised by the 29th World Medical Assembly in 1975 (2).

This document reads as follows:

INTRODUCTION

It is the mission of the medical doctor to safeguard the health of the people. His or her knowledge and conscience are dedicated to the fulfillment of this mission.

The Declaration of Geneva of the World Medical Association binds the doctor with the words, "The health of my patient will be my first consideration," and the International Code of Medical Ethics declares that, "Any act or advice which would weaken physical or mental resistance of a human being may be used only in his interest."

The purpose of biomedical research involving human subjects must be to improve diagnostic, therapeutic and prophylactic procedures and the understanding of the aetiology and pathogenesis of disease.

In current medical practice most diagnostic and prophylactic procedures involve hazards. This applies *a fortiori* to biomedical research.

Medical progress is based on research which ultimately must rest in part on experimentation involving human subjects.

In the field of biomedical research a fundamental distinction must be recognized between medical research in which the aim is essentially diagnostic or therapeutic for a patient, and medical research, the essential object of which is purely scientific and without direct diagnostic or therapeutic value to the person subjected to the research.

Special caution must be exercised in the conduct of research which may affect the environment, and the welfare of animals used for research must be respected.

Because it is essential that the results of laboratory experiments be applied to human beings to further scientific knowledge and to help suffering humanity, The World Medical Association has prepared the following recommendations as a guide to every doctor in biomedical research involving human subjects. They should be kept under review in the future. It must be stressed that the standards as drafted are only a guide to physicians all over the world. Doctors are not relieved from criminal, civil and ethical responsibilities under the laws of their own countries.

I. BASIC PRINCIPLES

1. Biomedical research involving human subjects must conform to generally accepted scientific principles and should be based on adequately performed laboratory and animal experimentation and on a thorough knowledge of the scientific literature.

2. The design and performance of each experimental procedure involving human subjects should be clearly formulated in an experimental protocol which should be transmitted to a specially appointed independent committee for consideration, comment and guidance.

3. Biomedical research involving human subjects should be conducted only by scientifically qualified persons and under the supervision of a clinically competent medical person. The responsibility for the human subject must always rest with a medically qualified person and never rest on the subject of the research, even though the subject has given his or her consent.

4. Biomedical research involving human subjects cannot legitimately be carried out unless the importance of the objective is in proportion to the inherent risk to the subject.

5. Every biomedical research project involving human subjects should be preceded by careful assessment of predictable risks in comparison with foreseeable benefits to the subject or to others. Concern for the interests of the subject must always prevail over the interests of science and society.

6. The right of the research subject to safeguard his or her integrity must always be respected. Every precaution should be taken to respect the privacy of the subject and to minimize the impact of the study on the subject's physical and mental integrity and on the personality of the subject.

7. Doctors should abstain from engaging in research projects involving human subjects unless they are satisfied that the hazards involved are believed to be predictable. Doctors should cease any investigation if the hazards are found to outweigh the potential benefits.

8. In publication of the results of his or her research, the doctor is obliged to preserve the accuracy of the results. Reports of experimentation not in accordance with the principles laid down in the Declaration should not be accepted for publication.

9. In any research on human beings, each potential subject must be adequately informed of the aims, methods, anticipated benefits and potential hazards of the study and the discomfort it may entail. He or she should be informed that he or she is at liberty to abstain from participation in the study and that he or she is free to withdraw his or her consent to participation at any time. The doctor should then obtain the subject's given informed consent, preferably in writing.

10. When obtaining informed consent for the research project, the doctor should be particularly cautious if the subject is in a dependent relationship to him or her or may consent under duress. In that case the informed consent should be obtained by a doctor who is not engaged in the investigation and who is completely independent of this official relationship.

11. In case of legal incompetence, informed consent should be obtained from the legal guardian in accordance with national legislation. Where physical or mental incapacity makes it impossible to obtain informed consent, or when the subject is a minor, permission from the responsible relative replaces that of the subject in accordance with national legislation.

12. The research protocol should always contain a statement of the ethical considerations involved and should indicate that the principles enunciated in the present Declaration are complied with.

II. MEDICAL RESEARCH COMBINED WITH PROFESSIONAL CARE (CLINICAL RESEARCH)

1. In the treatment of the sick person, the doctor must be free to use a new diagnostic and therapeutic measure, if in his or her judgment it offers hope of saving life, re-establishing health or alleviating suffering.
2. The potential benefits, hazards and discomfort of a new method should be weighed against the advantages of the best current diagnostic and therapeutic methods.
3. In any medical study, every patient—including those of a control group, if any—should be assured of the best proven diagnostic and therapeutic methods.
4. The refusal of the patient to participate in a study must never interfere with the doctor-patient relationship.
5. If the doctor considers it essential not to obtain informed consent, the specific reasons for this proposal should be stated in the experimental protocol for transmission to the independent committee (A, 2).
6. The doctor can combine medical research with professional care, the objective being the acquisition of new medical knowledge, only to the extent that medical research is justified by its potential diagnostic or therapeutic value for the patient.

III. NON-THERAPEUTIC BIOMEDICAL RESEARCH INVOLVING HUMAN SUBJECTS (NON-CLINICAL BIOMEDICAL RESEARCH)

1. In the purely scientific application of medical research carried out on a human being, it is the duty of the doctor to remain the protector of the life and health of that person on whom biomedical research is being carried out.
2. The subjects should be volunteers—either healthy persons or patients for whom the experimental design is not related to the patient's illness.
3. The Investigator or the investigating team should discontinue the research if in his/her or their judgment it may, if continued, be harmful to the individual.

4. In research on man, the interest of science and society should never take precedence over considerations related to the well-being of the subject.

With this as a background, we will proceed to guide the reader in the recruitment and use of human subjects.

PROTECTION OF HUMAN SUBJECTS IN RESEARCH

The use of human subjects in research has been a practice for thousands of years. Until relatively recently this practice has not been subject to any formalized regulatory process.

Compliance with the regulatory process and codes of practice is now necessary for certain types of human research, and the principles established in these codes and regulations have set the standards for research with human subjects for all other human subject research. Reading and understanding regulations regarding human subjects should be the first order of business for any person or group of persons who wish to engage in research employing human subjects.

For the purposes of this chapter the U.S. regulatory process is the underlying background for the information presented. Therefore, the *Federal Register* published regulations on the Protection of Human Subjects (3, pp. 8942–8980) should be consulted. The regulations were written for specific applications, but the principles applied in these regulations will guide the investigator to the general practices relative to the use of human subjects.

Prior to exploring the use of human subjects in research, the actual processes involved in carrying out human research should be detailed and defined.

Some key terms are defined as follows:

Clinical Investigation

A clinical investigation may be defined as any experiment that involves a test material and one or more human subjects. The test material, for our purposes, may or may not be subject to the requirements of the Food and Drug Administration. The reasons for carrying out the experiment or research project may be to seek new knowledge, to reevaluate existing knowledge, to verify theory, or to apply current knowledge to various situations.

Clinical investigations can be carried out solely for the benefit of the human subject; however, in many investigations the subjects may not receive any direct benefit from their participation in an investigation. When human subjects participate in investigations, they should understand the benefits and risks to be expected (i.e., improved health, beauty aid, allergic reaction).

In general, there are three typical categories of human research. One type of research is conducted to determine the safety of a product or procedures in humans. In this type of study the investigator is evaluating the human tolerance for the procedure or product. In drug evaluations this type of study is classified as a Phase I clinical evaluation.

The second category of human research is the efficacy phase. In these studies the product's efficacy for its intended use is determined, e.g., a Phase II clinical evaluation for a drug. The efficacy trials on products can range from a simple trial to many trials conducted with several investigators to determine wider margins of efficacy under multiple conditions and/or replications of the effects in various hands. The product type will usually determine the extent of these more extensive tests. In drug evaluations this would be Phase III.

In addition, the evaluation of products continues throughout the lifetime of a product or a drug. The consumer is always performing an ongoing evaluation of the efficacy or safety of a product. In this case the evaluation is not always conducted by a trained investigator, such as a physician evaluation of the performance of a drug in relation to a patient's responses. It can simply be the consumers' acceptance or rejection of the product as advertised for its intended use. In drug evaluations the continuous monitoring for performance and safety is termed Phase IV.

Because the pharmaceutical industry has a planned approach to the introduction of a safe and effective product to the market, it can be used as a model for other industries to follow. However, because the consequences of other product investigations are potentially less hazardous, the time factors and extent of testing with human subjects are considerably less in scope from the standpoint of time, cost, and numbers of subjects evaluated.

Human Subjects

A "human subject" can be defined as an individual who is a participant in a clinical investigation. The subject could be a normal, healthy volunteer or a patient volunteer. The human subject has been additionally defined by Brady and Jonsen as "a person about whom an investigator conducting research with the objective of developing generalized knowledge obtains the following: data through intervention of interaction with said person; and/or identifiable private information" (2, p. 9).

From the subject's perspective, then, two types of human research are distinguishable, therapeutic and nontherapeutic (4).

Minimal Risk

This term means that the risks of harm anticipated in the proposed research are not greater, considering probability and magnitude, than those ordinarily encountered in daily life or during the performance of routine physical or psychological examinations or tests (3, p. 8950).

If the elements of minimal risk are not present, the magnitude of risk is greater and therefore the gravity of the implications, both to the subject and to the acceptability of proceeding with the research project, must be considered.

Institutional Review Board

The regulations governing Institutional Review Boards (IRB) are presented in the *Federal Register* (3, pp. 8942—8980).

In summary, the IRB function is to assure that risks to human subjects are minimized and are reasonable in relation to anticipated benefits; that the selection of subjects is equitable and that each subject or his or her legally authorized representative will give their documentable informed consent; that the study plan will provide for monitoring the collected data to ensure the subject's safety; that the subject's right to privacy is protected and the data are confidential. In addition, the IRB must be sure there are appropriate safeguards to protect the rights of subjects who are members of vulnerable groups (i.e., children, prisoners, mentally impaired, etc.).

All human research is not governed by IRB review. The investigator or research sponsor has the obligation of assuring that the subjects are protected within the framework of the applicable regulations. Research projects not governed by the regulations should, however, be conducted within the spirit of the established codes and regulations.

The regulations have specified the minimum standards for the composition of the review board. All board members need not be present for a review but a quorum of the board must be present to take action. In some specific cases review can take place through expedited procedures and a single appointed member of the board may approve a project on behalf of the board.

The practice of most researchers suggests that informed consent would be an acceptable prerequisite for enrollment in human research.

Since the publication of IRB regulations in 1981, the Food and Drug Administration has attempted to keep sponsors of research, clinical investigators, and IRBs informed by conducting workshops and issuing information sheets, which attempt to clarify the issues surrounding the implementation of the regulations involving IRBs and informed consent. The current list of information sheets available from the Food and Drug Administration is shown in Table 1.

Informed Consent

The regulations governing informed consent are presented in the *Federal Register* (3, pp. 8942-8980). In summary, the informed consent of a subject who wishes to volunteer for research projects must be obtained prior to that subject's participation in a study. The consent in most cases should be a written document, but there are special circumstances where oral consent may be obtained. There may be rare occasions, such as life-threatening cases, where informed consent may be waived entirely.

The basic and additional elements of informed consent are presented in Table 2 (3, p. 8951).

There are special risk groups of potential subjects, such as children, mentally incompetent persons, and prisoners. The major obstacle in employing these subjects in research is the issue of obtaining informed consent. There are individual *Federal Register* publications on each of these groups.

The regulatory requirements for the use of human subjects place many obligations on the investigator. Understanding and implementing these regulations are only a fraction of the entire scope of performing research employing human subjects.

THE PROCESS OF PREPARING A PROGRAM
FOR HUMAN RESEARCH

Prior to exploring the use of human subjects in research, the actual processes involved in carrying out human research should be

Table 1 Food and Drug Administration Information Sheets

IRB Information Sheet: FDA Inspections

A Suggested Self-Evaluation Guide: Human Subject Protection Institutional Review Boards

Food and Drug Administration Regulations Which Relate to Institutional Review Board Activities

Waiver of IRB Requirements

Non-local IRB Review

Sponsor—Clinical Investigator—IRB Interrelationship

Clinical Investigators Unaffiliated with an Institution with an IRB

Significant Differences in HHS and FDA Regulations for IRBs and Informed Consent

Continuing Review

Cooperative Research

FDA District Offices

Investigational Use of Marketed Products

Emergency Use of a Test Article

Acceptance of Foreign Data and IRB and Informed Consent Requirements

Answers to Frequently Asked Questions

Payment for Investigational Products

Investigational Drug Use in Patients Entering a Second Institution

IRBs and Medical Devices

Guidance on Significant and Nonsignificant Risk Device Studies

Guidance for the Emergency Use of Unapproved Medical Devices

Payment of Research Subjects

Advertising for Study Subjects

Table 2 Basic and Additional Elements of Informed Consent

Basic elements of informed consent. In seeking informed consent, the following information shall be provided to each subject:

(1) A statement that the study involves research, and explanation of the purposes of the research and the expected duration of the subject's participation, a description of the procedures to be followed, and identification of any procedures which are experimental.

(2) A description of any reasonable foreseeable risks or discomforts to the subject.

(3) A description of any benefits to the subject or to others which may reasonably be expected from the research.

(4) A disclosure of appropriate alternative procedures or courses of treatment, if any, that might be advantageous to the subject.

(5) A statement describing the extent, if any, to which confidentiality of records identifying the subject will be maintained and that notes the possibility that the Food and Drug Administration may inspect the records.

(6) For research involving more than minimal risk, an explanation as to whether any compensation and an explanation as to whether any medical treatments are available if injury occurs and, if so, what they consist of, or where further information may be obtained.

(7) An explanation of whom to contact for answers to pertinent questions about the research and research subjects' rights, and whom to contact in the event of a research-related injury to the subject.

(8) A statement that participation is voluntary, that refusal to participate will involve no penalty or loss of benefits to which the subject is otherwise entitled, and that the subject may discontinue participation at any time without penalty or loss of benefits to which the subject is otherwise entitled.

Additional elements of informed consent: When appropriate, one or more of the following elements of information shall also be provided to each subject:

(1) A statement that the particular treatment or procedure may involve risks to the subject (or to the embryo or fetus, if the subject is or may become pregnant) which are currently unforeseeable.

(continued)

Table 2 (Cont.)

(2) Anticipated circumstances under which the subject's participation may be terminated by the investigator without regard to the subject's consent.

(3) Any additional costs to the subject that may result from participation in the research.

(4) The consequences of a subject's decision to withdraw from the research and procedures for orderly termination of participation by the subject.

(5) A statement that significant new findings developed during the course of the research which may relate to the subject's willingness to continue participation will be provided to the subject.

(6) The approximate number of subjects involved in the study.

The informed consent requirements in these regulations are not intended to pre-empt any applicable Federal, State, or local laws which require additional information to be disclosed for informed consent to be legally effective.

Nothing in these regulations is intended to limit the authority of a physician to provide emergency medical care to the extent the physician is permitted to do so under applicable Federal, State, or local law.

---

detailed. The preparation of a protocol that will be the guiding document for the research project is the first step. However, before human work is undertaken, a substantial/reasonable amount of preclinical data should be available to satisfy the investigator that the test materials possess an acceptable level of safety in order to proceed into humans.

Table 3 presents a scheme of the preparation process for implementing a research project.

The Protocol

Before any research project can be initiated, there should be the reasoning process that asks, "What is the question for which we

Table 3 Process for Implementation of a Research Project

1. Prepare protocol.
2. Assemble safety data on test products—preclinical.
3. Investigator evaluates protocol and safety data.
4. Investigator agrees to carry out study.
5. Investigator prepares IRB documentation.
   a. Protocol (and advertisement, if to be used).
   b. Safety issues regarding project.
   c. Informed consent statement and payment schedule.
   d. Subject instructions.
6. Investigator receives IRB approval.
7. Proceed with study.
   a. Schedule of events.
   b. Recruit study.
      1) Screen subjects.
      2) Select participants.
      3) Carry out procedures—document events and data.
      4) Adverse reaction—notification of sponsor and IRB (regulations).
      5) Subject follow-up.

are seeking answers?" When there is a clear idea of the question, then the methods to seek a proper resolution of that question can be formulated.

A number of final and tentative final monographs are available. A working knowledge of these monographs can be invaluable in planning the study and in preparation of the protocol. It is also a useful aid in determining the numbers and types of test panels to be selected for a given research program.

The essence of an investigation involving human subjects should be the study protocol or plan. The protocol must contain the following items:

1. Objective: the reason for carrying out the study—a clear statement of the goal to be achieved
2. The materials to be tested
3. The procedures to be used to carry out the investigation

4. The parameters to be observed or measured as well as the reporting documents on which the data will be recorded
5. The number and type of human panelists to be used—details of the inclusion and exclusion criteria
6. The methods to be used in analyzing the data

When the above items are all put forth in the protocol, the person responsible for carrying out the project, the investigator, can determine whether the project is feasible and whether there are available subjects who can participate in such a study. In addition to subject availability, the investigator must determine the potential for risk or benefit to the participants.

Once these are known the investigator can then prepare an informed consent statement, which will be necessary for the human subject to sign prior to entering the research project. Not all study types may require informed consent, but the investigator should make himself aware of these instances. Here again, the guidelines for such a decision should be a working knowledge of the Protection of Human Subjects regulations.

In addition, the regulations provide specific definitions of certain key words, which allows the investigator to have a better understanding of the scope of the regulation in relation to the defined terms.

Investigational New Drug (IND): A drug product that is not yet approved for marketing in the United States.

New Drug Application (NDA): An application that is submitted to the Food and Drug Administration requesting approval to market a drug product. It includes the toxicological and clinical data that the sponsor wishes to submit to support the request.

Form FD 1571—Notice of Claimed Investigational Exemption for New Drug (Attachment 1): This form is completed and submitted to the Food and Drug Administration by the sponsor. It is the sponsor's request to use an investigational drug in clinical trials. It carries a 30-day waiting period before trial can begin, and the Food and Drug Administration may disallow the request any time within that 30-day period.

Form FD 1572—Statement of Investigator (Attachment 2): This form is completed by the investigator, and forwarded to the sponsor. It is a statement of the investigator's qualifications, institution affiliation, and his responsibilities in regard to peer review (IRB), drug disposition, and conduct of the study. It is used for Phase I studies and all research involving normal, healthy subjects.

Form FD 1573—Statement of Investigator (Attachment 3): This is the same as Form FD 1572 except that it is used for Phase III studies involving subjects having the condition to be treated/prevented by the investigational drug.

Form FD 1639—Drug Experience Report (Attachment 4): This form is completed by the investigator and the sponsor in the case of serious adverse reaction in a subject participating in a clinical trial.

Investigator's Brochure: This document is a synopsis of all toxicological and clinical data to date on the investigational drug. It is prepared by the sponsor and a copy is sent to the investigator for his or her use and the use of the IRB in judging the safety of the proposed project. It is a confidential document.

While all of the above apply specifically to drug research, the logic of the processes used to control the human experiment makes sense. Even though human subject research carried out on non-drug test materials and/or procedures need not require all of the above documents, the rationale for conducting the research should follow the same basic process.

SUMMARY

The use of human subjects in research is a necessity in today's environment where the public is demanding that safe and efficacious products be available in the marketplace. The world body has expressed its desire to protect human subjects from exploitation through international and national codes of practice.

Sponsors of human research as well as investigators employing human subjects must keep abreast of all the principles and regulatory codes that are available.

REFERENCES

1. Brady, J. V., and Jonsen, A. R. The evolution of regulatory influences on research with human subjects. In: *Human Subjects Research; A Handbook for Institutional Review Boards* (Robert Greenwald, ed.). Plenum Press, New York (1982).

2. Declaration of Helsinki. *Med. J. Aust.* 1:206–207 (1976).

3. Protection of Human Subjects; Informed Consent; Standards for Institutional Review Boards for Clinical Investigations; and Clinical Investigations Which May Be Reviewed Through

Expedited Review Procedure. *Fed. Register* 46(17) (Tuesday, January 27, 1981).

4. Walters, Le Roy. Some ethical issues in research involving human subjects. *Perspect. Biol. Med.* p. 193 (Winter 1977).

2
Dermatotoxicology: Animal Irritancy Testing

SUSAN S. ROPER* *Emory University School of Medicine,*
Atlanta, Georgia

INTRODUCTION

To date dermatotoxicological research has been dominated by industrial specialists in formulating and testing new household detergents, industrial cleaners, and cosmetics for human use. Each major product company involved in these areas employs its own dermatotoxicological specialists or dermatological consultants to monitor animal and human irritancy testing programs. Federal and international regulations recommend that animal irritancy and toxicity studies be performed on chemicals that may contact human skin, including skin and eye irritation testing as well as evaluation of systemic toxicity from dermal absorption of the product (1-4). From these studies has accumulated much of our experience with testing of irritants in animals, which has been utilized principally as a screening tool for human irritancy studies.

**Present affiliation:* University of South Florida School of Medicine, Tampa, Florida.

This chapter presents to the neophyte researcher, unfamiliar with industrial dermatotoxicology, some of the general principles and techniques of animal irritancy testing. Predictive animal irritancy testing, as devised by the industrial dermatotoxicologists, is concerned principally with the potential of a product to irritate in one of four ways: acute primary irritation, cumulative irritation, corrosion, and phototoxic irritation (5,6). Acute primary irritation is defined as reversible, nonimmunological inflammation resulting from one application of the product. Cumulative irritation is defined as reversible inflammation resulting from repeated exposure to the product. Neither acute nor cumulative irritation, as defined, should result in scarring. Corrosion is used to describe skin necrosis resulting from one application of the product that results in scar tissue. Phototoxic irritation is defined as irritation resulting from photo-induced changes in the test product. Animal photoxicity models are not discussed here.

THE DRAIZE-FHSA TEST

Draize et al., in 1944, developed an animal model for measuring acute primary irritation, which has since been adapted by the Federal Hazardous Substance Act (FHSA) and is outlined in the Code of Federal Regulations (1500.41) (3,8). Draize also developed the procedure for assaying dermal systemic toxicity, which has also been adapted by the FHSA (CFR1500.40) (2). Dermal systemic toxicity studies refer to the LD_{50} of a chemical or product—that concentration or dose which is fatal in 50% of the test animals when applied to 10—30% of the skin—will not be discussed here. The Draize-FHSA procedure for assaying primary irritation is as follows:

Primary irritation to the skin is measured by a patch test technique on the abraded and intact skin of the albino rabbit, clipped free of hair. A minimum of six subjects are used in abraded and intact skin tests. Introduce under a square patch, such as surgical gauze measuring 1 in. by 1 in. and two single layers thick, 0.5 ml (in the case of liquids) or 0.5 g (in the case of solids and semisolids) of the test substance. Dissolve solids in an appropriate solvent and apply the solution as for liquids. The entire trunk of the animal is then wrapped with an impervious material, such as rubberized cloth, for the 24-hr period of exposure. This material aids in maintaining the test patches in position and retards the evaporation of volatile substances. After 24 hr of exposure, the patches are removed and the resulting reactions are evaluated on the basis of the designated values (Table 1). Readings are

Table 1 Evaluation of Skin Reaction

Skin reaction	Value[a]
Erythema and eschar formation	
No erythema	0
Very slight erythema (barely perceptible)	1
Well-defined erythema	2
Moderate to severe erythema	3
Severe erythema (beet redness) to slight eschar formations (injuries in depth)	4
Edema formation	
No edema	0
Very slight edema (barely perceptible)	1
Slight edema (edges of area well defined by definite raising)	2
Moderate edema (raised approx. 1 mm)	3
Severe edema (raised more than 1 mm and extending beyond the area of exposure)	4

[a]The "value" recorded for each reading is the average value of the six or more animals subject to the test.

again made at the end of a total of 72 hr (48 hr after the first reading). An equal number of exposures are made on areas of skin that have been previously abraded. The abrasions are minor incisions through the stratum corneum, but not sufficiently deep to disturb the dermis or to produce bleeding. Evaluate the reactions of the abraded skin at 24 hr and 72 hr, as described here. Add the values for erythema and eschar formation at 24 hr and at 72 hr for intact skin to the values on abraded skin at 24 hr and at 72 hr (four values). Similarly, add the values for edema formation at 24 hr and at 72 hr for intact and abraded skin (four values). The total of the eight values is divided by four to give the primary irritation score.

The Draize-FHSA procedure, or some modification of it, has been utilized by many industrial dermatotoxicology laboratories for screening potential human irritants. Many, however, have criticized the procedure as being too sensitive and have developed modifications of the procedure to suit their testing programs. In addition to the private laboratories, national and international agencies have modified the Draize-FHSA procedure, as shown in

Table 2 (5,9−11). The Organization for Economic Cooperation and Development, an international agency, published in 1981 the OECD Guidelines for Testing of Chemicals. The portion of the guidelines dealing with skin irritation testing in animals is reprinted with permission in the appendix. The different variables of the Draize-FHSA procedure will be discussed separately.

Laboratory Animal Species

The New Zealand albino rabbit, with its highly sensitive skin, is recommended for standard Draize-FHSA testing and in the OECD Guidelines (3,5,9,12). The rabbit is the most popular laboratory animal species used. Some researchers, however, prefer the Hartley albino guinea pig because of its smaller size, lower cost, and ease of handling (1,13−15). The National Academy of Sciences favors using the guinea pig over the rabbit as the guinea pig's skin is less sensitive than the rabbit's and is closer in sensitivity to human skin (1,16). Guinea pigs also are the animal model used most often for skin allergic contact sensitization studies. Most laboratories, however, utilize more than one laboratory animal species to test their products before testing on humans. This is specifically recommended in the OECD Test Guidelines (12). In addition to rabbits and guinea pigs, hairless mice, rats, mice, and swine have been used in animal irritancy studies (17−24). Hairless rats and hamsters in exploratory studies have shown little capacity to respond to skin irritants. Table 3 depicts the species in order of tested skin sensitivity (16,19,22, 23,26).

Differences in skin sensitivity are species specific and may be due to differences in skin permeability. To investigate this, Bartek et al. used six radioactively labeled compounds applied to the skin of the rat, rabbit, pig, and human. The rabbit's skin was shown to be up to 10 times more permeable than human skin (27).

The marked sensitivity of rabbit and guinea pig skin has led to numerous complaints and doubts concerning the predictive value of animal irritancy studies (20,23). Kastner reported on the wide variability of irritant reactions between humans and laboratory animals to the same product (23). Campbell and Bruce (28) and Motoyoshi et al. (19), in separate studies, found that isopropyl myristate, a ubiquitous component of cosmetics and topical medicaments, is a moderate to severe irritant in rabbits at the same concentration that is minimal to negative in humans. The rabbit's skin seems to be especially sensitive to the class of fatty-

Table 2 Comparison of Regulatory Requirements

Source	Test material		Exposure time (hr)	Number of rabbits	Sites per animal (intact/abraded)	Action at end of exposure	Scoring periods after exposure
	Solid	Liquid					
OECD, 1981	Moisten	Undiluted	4	3[a]	1/0	Wash with water or solvent	30–60 min, 24, 48, 74 hr, or until obviously irreversible; no longer than 14 days
FHSA (CFR, 1985)	Dissolve	Neat	24	6	1/1	Not specified	24 and 72 hr
DOT (CFR, 1985)	Not specified	Not specified	4	6	1/0	Wash with appropriate solvent	4 and 48 hr

[a]Additional animals may be required to clarify equivocal results.
Source: McCreesh and Steinberg (5), reprinted with permission.

Table 3 Comparison of Skin Sensitivity of Laboratory Animal Species

	Most sensitive	Least sensitive
Davies (1972) (22)	Rabbit, mouse, piglet, guinea pig, dog, human, minipig, baboon	
MacMillan (1975) (26)	Rabbit, guinea pig, human, beagle dog	
Kastner (1977) (23)	Rabbit, guinea pig, hairless mice, human	
Motoyoshi (1979) (19)	Rabbit, guinea pig, albino rats, human, pigs	
Griffith and Buehler (1977) (16)	Rabbit, guinea pig, human	

acid-derived compounds, which seldom irritate human skin (10, 19). These interspecies differences emphasize the need for human irritancy studies before rejecting or marketing a product for human use.

Immobilization

Immobilization of the animal during the testing procedure, after the irritants are applied, may be unnecessary, as it increases undue stress on the animals. It is specifically not recommended by the OECD Guidelines (12). Different laboratories may vary on the need for immobilization. Some laboratories utilize Elizabethan collars (12,28) or leather harnesses (26) to prevent the animals from removing the patches once they are applied.

Clipping the Fur

Twenty-four hours prior to application of the irritants, the hair of the dorsal side of the animal is clipped with an animal clipper (Oster), taking care not to abrade the skin. Some researchers prefer to epilate the hair after clipping with barium sulfide (Magic Shave) or Neet (13). The advantage of epilating is to enhance the visibility of mild irritant reactions, which can be difficult to read on nonepilated animal skin. Epilation must be performed 24 hr prior to applying the irritants, and if done correctly, it does not visibly irritate the skin. The disadvantage of epilation is that it is messy, time-consuming, and may subclinically disturb the barrier function of the skin.

Abrasion

Recommended test procedures differ on the need to test on abraded skin (Table 1). It is specifically recommended in the Draize-FHSA procedure (3,8), whereas it is specifically not recommended in the OECD Test Guidelines (12). McCreesh and Steinberg (5), Nixon et al. (29), and Roudabush et al. (13) agree that testing on abraded skin can be misleading in the interpretation of test results and may be unnecessary. Marzulli and Maibach feel that both intact and abraded skin sites should be tested (25). Kligman and associates feel that utilization of abraded skin techniques is invaluable in human irritancy studies (27,30). The most popular method of abrasion is to use the tip of a sterile needle to make four (3,8) to eight (30) crosshatch marks on the skin. Care should be taken that only the stratum corneum is abraded and that no bleeding results. One may also use the Berkeley Scarifier (Berkeley Biologicals, Berkeley, CA) or the Maryland Plastics skin abrader (Maryland Plastics, Federalsburg, MD). McCreesh and Steinberg have demonstrated the equivalency of these abraders with the needle technique (5,6).

Patch Test Material

The Draize-FHSA procedure calls for a 1-in.-square semiocclusive patch made of gauze. Materials used have included cellulose pads (13), surgical gauze (3,31–33), Melolin (34), Webril pads (Kendall) (26,30,35), and cotton flannel pads (17,18). Adhesive tape is used to hold the patches to the animal's skin including Blenderm (3M) (28,26), Dermicel (Johnson and Johnson) (28), and Elastoplast (Beiersdorf, Inc.) (28).

Number of Patches

The Draize method recommends four patches per rabbit—two intact and two abraded (8). The guinea pig, being smaller, accommodates two patches per animal. In the FHSA guidelines, only two patches per animal are applied—one intact and one abraded (3). The OECD Test Guidelines recommend only one patch per animal, without abrasion (12). Most laboratories using the rabbit prefer to use four patches per animal, similar to what Draize originally recommended.

Placement of Patches

The placement of the patch tests should be standardized to compare results and to control for regional variability in skin sensitivity (36). It is well known that the less hirsute abdominal region of the animal is more irritable than the clipped hirsute dorsal side of the animal, which is the recommended test area in the Draize-FHSA procedure. The incomplete Latin square or block design with rotation of test sites has been recommended to minimize site-to-site and animal-to-animal variability (19,28,32). Owing to the small size of the test area, this may be unnecessary.

Method of Application

The test substance, according to the Draize-FHSA procedure, is to be inserted under the patch after it has been taped to the skin. Some laboratories vary by applying the material directly to the skin before applying the patch, or by applying the material directly to the patch before taping the patch to the skin (17, 20, 34). Applying the substance to the patch first may decrease the amount of material in contact with the skin, although Steinberg et al. have found no difference between applying the material to the skin site or to the patch first (37). The most popular method to apply the test substance is to inject it with a needle and syringe through the gauze, which has already been taped to the skin (31,32). Some researchers recommend that liquids containing volatile solvents should first be applied to the skin and allowed to evaporate before occluding, in order to eliminate the inherent irritancy of the solvent (1,28,38).

Dosage and Size of Test Area

The quantity of test substance used may vary with the size of the test area or amount of the test substance available. The Draize-FHSA procedure calls for 0.5 g or 0.5 ml over a 1-in.^{2} (2.5 cm^2) area (6.25 cm^2) (3,8). The OECD Test Guidelines recommends 0.5 g or 0.5 ml over an approximately 6 cm^2 area (12). In actuality, the test site has varied from 1 cm^2 to 2 in.^{2} with 1 in.^{2} and 2 cm^2 being the most popular. The size of the test dose has varied from 10 µl to 1 ml. A micropipet is recommended as an excellent tool for accurately measuring the dose of test material.

Occlusion

The occlusivity of the patch test is an important variable that differs with the patch test materials used, the tape, and the method of covering the patches. Holland et al. (39) and Fernstrom (40) have both reported studies in human volunteers demonstrating the heightened irritancy of occluded patch tests. The Draize-FHSA procedure utilizes a semiocclusive patch, which is taped to the skin and covered with an "impermeable" material such as rubberized cloth. Steinberg et al. use Elastoplast covered with a stockinette sleeve, which becomes an occlusive dressing (5,37). Gilman et al. found that patches covered with Elastoplast were significantly more irritated than patches not covered with Elastoplast (33). Steinberg et al. found that cellophane and Elastoplast occlusion are equivalent methods, yielding similar test results in animal irritation studies (37). The OECD Test Guidelines recommend the use of a semiocclusive patch loosely taped to the skin, stating that occlusive tests are too "severe" (12). Occlusive vs. semiocclusive patch testing in animals has not been clearly defined or interpreted by either the research laboratories or the recommending guidelines.

Increasing the pressure evenly over the patch test enhances the occlusivity as well as the irritant reactions (40-42). Uneven pressure on a patch, however, may lead to more severe reactions on the edge of the test area secondary to uneven distribution of the test substance (43). Certain patch tests may cause irritation themselves, when occluded. Precautions should be taken to eliminate or control for this factor (5).

It may be more feasible in certain situations to use an open patch test system. The effect of patch test materials and occlusivity is eliminated. Volatile solvents evaporate quickly in open test systems and are less irritating than if occluded (1,28). If the test is carefully performed, the data are reproducible (19, 44). Some investigators prefer open patch test systems for cumulative irritancy studies because of the closer approximation to human use situations (9,25,26,45,46). Open patch tests are also used more frequently in the non-Draize animal irritancy studies, discussed later. Some investigators, however, feel that the closed patch test yields more consistent results than the open patch test method (5,9). For a more meaningful assessment of acute primary irritation, both open and closed patch test systems may be necessary (11). To the contrary, some substances have been found to be even less irritating when occluded as compared to open testing, although the reverse is usually true (10,54).

Testing Dry Compounds

Methods of testing dry compounds may vary. The Draize-FHSA procedure suggests dissolving 0.5 g of solid in an appropriate vehicle and applying it under the patch. This may add an unknown percutaneous absorption factor, which should be controlled for with solvent controls. The amount of solid used, however, is usually less than 0.5 g. Gilman et al. showed that testing a dry product as a paste (83% concentration) or as a 50% suspension yielded essentially equivalent results if occluded for only 24 hr (33). Sullivan et al. compared five different methods of testing dry detergents in the Draize-FHSA procedure and found variation in results indicative of the manner in which the material was applied to the skin (48).

Vehicles

The vehicles used should be standardized to eliminate any differences in percutaneous penetration and inherent irritancy. Each researcher must decide how to test materials containing volatile solvents—whether or not to allow the material to evaporate before occluding. Ethanol is nonirritating if it is allowed to evaporate from the patch site before occlusion (28,38). Acetone, when occluded, is an irritant, and it should be diluted or avoided, if possible. In humans, differences in irritancy between occluded patch tests and usage tests have been attributed to the irritancy of the occluded volatile solvents (9). Campbell and Bruce tested corn oil, petrolatum, and two grades of mineral oil on rabbit skin and found that all, if occluded long enough, were irritants. Petrolatum, however, is nonirritating if only occluded for 24 hr. It is widely used as a vehicle in human patch test studies. Aqueous solutions should be tested with a deionized distilled water control (28).

Exposure Time

The length of time the material is in contact with the skin should be standardized, as irritancy may vary with exposure time. The Draize-FHSA procedure recommends a 24-hr exposure time, with readings taken at 24 and 72 hr (3). The OECD Test Guidelines (12) and the Department of Transportation guidelines (11) recommend shortening the exposure time to 4 hr in order to more realistically approximate human exposure to chemicals not intended for prolonged contact with human skin. The OECD Test Guidelines recommend readings to be taken at 4-1/2 to 5, 24, 48, and 72 hr after exposure. Most laboratories utilize the 24-hr exposure time recommended in the Draize-FHSA procedure (25,30).

The treatment of the skin test site after the patch is removed may also vary. The Draize-FHSA procedure does not specify cleansing of the test site after patch removal, whereas it is specifically recommended in the OECD Test Guidelines (3,12). The effect of this variable has not been adequately studied.

Scoring Reactions

As many different scoring systems exist as there are dermatotoxicologists. Most laboratories doing Draize-FHSA testing utilize the Draize-FHSA scoring system published in the Code of Federal Regulations (1500.41) and reprinted in Table 1. The reactions are read at 24 and 72 hr and are given two scores each--one for erythema/eschar and one for edema. The 24- and 72-hr readings for the intact and abraded skin sites yield eight values, which are summed and divided by 4 to yield the Primary Irritation Score. Different laboratories may use the same table but process the data differently, depending on the experimental design. Some laboratories prefer to measure edema with sensitive skin calipers (Schnelltester or Lange) instead of using the subjective Draize-FHSA scoring system (25). The reading and scoring of the irritant skin reactions is the most subjective component of the entire Draize-FHSA procedure (5,9,31). Individual bias in reading reactions accounted for much of the interlaboratory variability found by Weil and Scala in a study involving 25 laboratories and testing 12 chemicals (31). They concluded that regular clinics in methodology should be required for the technicians in order to standardize the reading of reactions. Steinberg et al. discovered that technicians tend to rate reactions more strongly than their toxicologist or veterinarian colleagues (37). At the present time, therefore, no reliable reference test exists that will yield reproducible interlaboratory toxicological data.

Some investigators have used vital dyes, such as trypan or sulfan blue, to enhance the visibility of mild inflammatory reactions (17–20,49). The value of these dye extravasation techniques over visual assessment of erythema is questionable (9).

Other investigators have done extensive histological assessments of the inflammatory reactions of the irritant reactions in order to objectively corroborate the more subjective visual scoring system (19,25,34,44,49,50). This work has also been done in humans (51). Clinically, as well as histologically, it is relatively easy to distinguish mild from severe irritant reactions. Mild to moderate reactions, however, are not as easily separated histologically. With early inflammation, the first change seen is minimal hyperkeratosis and acanthosis. Following this a mild inflammatory infiltrate appears. With greater inflammation, the degree of hyperkeratosis, acanthosis, and infiltrate increases. Dermal edema and spongiosis also appear as inflammation increases. The number of animals required, cost, and time these histological assessments require make this an impractical procedure for most animal irritancy studies.

More objective methods of measuring skin irritancy have been devised. An Evaporimeter (ServoMed, Sweden) is now being utilized to assess the irritancy of chemicals on animal and human skin as a function of transpidermal water loss (TEWL), which is a measurement of epidermal barrier dysfunction. Pitts et al. found that in measuring TEWL in guinea pig skin pretreated with irritating surfactants, the TEWL paralleled quantitation of visible erythema and was an even more sensitive indicator of barrier dysfunction than erythema (52). Van der Valk et al., in using the Evaporimeter to test irritants on human skin, found a 0.98 correlation between visual assessment of erythema and measured TEWL

(53). Thus, the Evaporimeter may become a valuable tool for generating parametric reproducible data in skin irritancy studies.

The Laser-Doppler Flowmeter (Periflux, PeriMed, Sweden) or Velocimeter (LD 5000, MedPacific, Seattle, WA) is another research tool being utilized by investigators in Denmark and San Francisco to replace the 133 Xenon washout technique as a simpler method of noninvasively measuring cutaneous blood flow (54–56). It is most sensitive in measuring erythema. Monochromatic light from the laser is transmitted to the skin through an optical fiber to a depth of approximately 1 mm over an area of 2– 4 mm^2. The light is back-scattered with Doppler-shifted frequencies, as it is reflected from the moving erythrocytes in the cutaneous vasculature, and with unshifted frequencies from the surrounding stationary tissue. The reflected frequencies are collected by an optical fiber system, processed, and quantitated. This method has been shown in human patch testing to be an effective tool in separating negative and doubtful from positive patch test responses (57). The Laser-Doppler Flowmeter may be adaptable in the future to quantifying erythema in animal irritancy studies and eliminating the interlaboratory variability in dermatotoxicological test results found by Weil and Scala (31).

Controls

Each animal serves as its own control. Solvent controls, as mentioned previously, are necessary for both aqueous and nonaqueous test solutions. Some investigators also prefer to include positive irritancy controls, such as sodium lauryl sulfate (9,17). Positive irritant controls, if used, should be performed on separate animals in order to eliminate the initiation of conditioned hyperirritability (58). In guinea pigs, it has been demonstrated that one small area of inflammation 1 cm in diameter heightens the irritability of the surrounding normal skin to mild irritants (44). In humans, Bjornberg has demonstrated the heightened irritability of normal skin in patients with hand eczema (59).

Statistics

Statistical techniques for nonparametric data are needed. These include the least-significant-difference technique of Cochran and Cox (60), the signed rank sum test of Wilcoxon et al. (61), and the Spearman rank correlation coefficient technique (62).

Criticism

Most of the flaws of the Draize-FHSA procedure have been detailed, including the subjectivity of the scoring procedure, the variability of occlusive techniques, and the unusual sensitivity of rabbit skin to solvents, hydrocarbons, and fatty acid derivatives (10). In order to eliminate these problems, some investigators have developed other methods of animal irritancy testing, which will be discussed next.

OTHER METHODS

Guinea Pig Immersion Test

The guinea pig immersion test is used by a number of industrial laboratories of cosmetic and soap manufacturers to provide consistent, sensitive, and reproducible data on primary cutaneous irritation from water-soluble materials (14,26,63,64). F. H. Snyder and D. L. Opdyke have been credited with the original design of this method (26). Opdyke, with over 20 years of experience with this method, has stated that this procedure is 100% effective in screening products for potential human irritant reactions (14). In this procedure, unclipped guinea pigs are immersed in a 40°C bath of the test solution for 4 hr, rinsed, and dried, daily for 3 days. Two days following the third immersion, the animals are clipped and the irritant reactions on the abdominal skin are graded on a scale of 1 to 10. If the solution is judged to be nonirritating, the concentration of the test solution is doubled until an effect is produced (14). Pitts et al. have used the guinea pig immersion test in evaluating the effect of environmental variables on TEWL. They have also used it as a method to create an animal model with barrier-deficient skin on which to test their products (52). The main criticism of this test is that it cannot be used to test insoluble or nondispersible materials.

Formalin—Trypan Blue Test

Finkelstein et al. advocate and use a formalin—trypan blue test procedure to evaluate mild irritants (17,18). They have used rats, rabbits, and guinea pigs in their studies with equal success. In this procedure, three applications of 20% formaldehyde are painted on the test sites, which are located on the abdominal side of the anesthetized animal. Four to eight test sites per animal are used, depending on the species used (four for rat, six for guinea pig, and eight for rabbit). The test substance is impregnated on a 1-in. circular flannel pad and applied to the test site.

Finkelstein et al. also use a positive control with a substance of known irritancy. A sheet of polyethylene is placed over the entire abdominal area for occlusion. One-half to two milliliters of 0.5% trypan blue is injected into the axilla of the animal. After 16 hr the polyethylene sheet and the patches are removed and the degree of irritation is quantitated by scoring the intensity of the blue color on a scale from 0 to 100%. Finkelstein has stated that this method is much more sensitive in his hands than the Draize-FHSA method for distinguishing mild from moderate irritants, and that it correlates well with his 4-day repetitive occlusive human patch test technique. Others, however, have been unable to reproduce his work (20).

Mouse Ear Test

Uttley and van Abbe reported in 1973 on their success with the mouse ear test as an excellent method for screening human irritants before testing in humans (21). They used CF1 female albino mice in groups of five for each substance to be tested. On 4 successive days, 10 µl of liquid or 10 mg of solid in cream form is applied to the dorsal aspect of one ear, the other ear serving as a control. Reactions are read 24 hr after the fourth application on a scale of 0 to 14 and averaged for the group. They do not rely on this test exclusively, however, and recommend using more than one species for animal irritancy testing. They supplement their mouse data with testing in rabbits.

Repetitive Patch Test Techniques

Repetitive patch test techniques for animal irritancy studies (6, 20, 24–26, 34, 37, 46, 50, 65, 66) have been devised in parallel with human repetitive patch test techniques (18, 24–26, 30, 32, 35, 47, 50, 67) to assess the irritancy of chemicals contained in medicaments, cosmetics, and toiletries intended for frequent human use. Many repetitive techniques utilize open patch testing systems (25). Repetitive patch testing is a severe test of the phenomenon known as "skin fatigue," wherein the skin reacts to an accumulation of insults at subthreshold intensity (65, 68). One problem associated with repetitive patch testing is the appearance of allergic contact reactions in rabbits using test materials not known to be sensitizers (66). Repetitive patch testing may also result in "hardening" or "accommodation" of the skin, wherein the skin becomes resistant to further irritation, reacting less intensely with each irritant application (68). Differences in exposure times, occlusivity, and treatment period make comparison of the various repetitive techniques impractical for this discussion.

Lanman et al. found the 21-day human continuous closed patch test to be an extremely sensitive discriminator of human irritants (47). Steinberg et al. demonstrated that a 21-day closed continuous patch test in rabbits using irritants at threshold concentrations offered excellent predictability of the human repetitive patch test (37). Some human irritants, however, do not react well on rabbit skin, even though the rabbit's skin is much more sensitive than human skin. For example, formaldehyde is much more irritating in humans than rabbits in the closed patch test; however, it becomes an irritant in rabbits in the open patch test (37). Other investigators have identified irritants with similar properties in animal testing systems (23,29). This emphasizes that animal irritancy testing cannot be relied on exclusively to identify potentially hazardous human irritants.

Human Irritancy Testing

Although the scope of this chapter does not include human irritancy testing, a brief review is in order. In 1944, Schwartz and Peck described the "prophetic patch test" as a method to screen potential irritants and sensitizers in the product market (69). Various modifications of this patch test, in combination with usage tests such as the human arm immersion test (24), have been used over the years to assess the irritancy of products marketed for human use. Kligman and Wooding introduced the concept of the ID_{50}—the irritant concentration needed to produce a discernible reaction in 50% of the population tested in 24 hr—and the IT_{50}—the estimated number of days of continuous exposure that will produce a threshold reaction in 50% of the population tested (67). Lanman et al. touted the 21-day occlusive repetitive human patch test as the most sensitive predictor of human irritancy reactions (47). Frosch and Kligman, however, abandoned this technique after encountering irritant reactions in individuals which were not predicted by this method (30). They devised the "chamber-scarification" test, a 3-day repetitive occlusive patch test on abraded skin utilizing the "Duhring soap chamber," an 8-mm-diameter aluminum patch similar to the Finn Chamber, except with a deeper well. More recently, however, Mills et al. have shown that their chamber-scarification test, although a useful tool in screening human irritants, is not as sensitive as usage testing in predicting irritation (70).

CONCLUSION

The major flaws of animal irritancy testing involve the difficulty of reproducing results and standardizing techniques among laboratories (31). Intralaboratory variability, however, has been minimized, as each laboratory has refined its own individual techniques for predictive animal irritancy testing. The major industrial laboratories have used animal irritancy studies as a means to an end—marketing a safe product for human use—and have been satisfied with their techniques as long as this goal is met. It becomes difficult, therefore, to extrapolate and compare data among laboratories making much of the accumulated knowledge uninterpretable, especially to the nonindustrial researcher. Nevertheless, the subjectivity of the recording techniques will be replaced in time with more objective methods, discussed here, which should eliminate much of the interlaboratory variability. Until then, each investigator should be principally concerned with the reproducibility of data within his or her own laboratory.

APPENDIX: OECD GUIDELINES FOR TESTING OF CHEMICALS*

Preface

General

1. This publication contains the official OECD[†] guidelines for the Testing of Chemicals as adopted by the OECD council.

2. The Test Guidelines have been developed initially under the OECD Chemicals Testing Programme (see paragraphs 10–16 below), and subsequently since 1981, as provided by the council under the OECD Updating Programme for Test Guidelines.

3. Whenever testing of chemicals is contemplated, the OECD Test Guidelines should be consulted. Since the Test Guidelines have been endorsed by the OECD Member countries, their use in the generation of data provide a common basis for the acceptance of data internationally, together with the opportunity to

---

*The following portions of the OECD Guidelines for Testing of Chemicals reprinted with permission.
†Organization for Economic Cooperation and Development.

reduce direct and indirect costs to governments and industry associated with testing and assessment of chemicals.

4. Other methods and guidelines not included in this publication may be judged to be appropriate in testing chemicals in certain scientific, legal, and administrative contexts.

5. The OECD Council Decision on Mutual Acceptance of Data (12th May, 1981; C(81)30) affirms that data generated in one country in accordance with the OECD Test Guidelines—and additionally in accordance with the OECD Principles of Good Laboratory Practice—should be accepted in OECD countries for purposes of assessment and other uses relating to protection of man and the environment. The full text of this Decision and the OECD Principles of Good Laboratory Practice may be found in the Appendix to the OECD Guidelines for Testing of Chemicals.

6. The OECD Test Guidelines contain generally formulated procedures for the laboratory testing of a property or effect deemed important for the evaluation of health and environmental hazards of a chemical. The Guidelines vary somewhat in respect of detail, but include all the essential elements which, assuming good laboratory practice, should enable an operator to carry out the required test.

7. OECD Test Guidelines are not designed to serve as rigid test protocols. They are instead designed to allow flexibility for expert judgment and adjustment to new developments.

8. It is intended that the OECD Test Guidelines be used by experienced laboratory staff familiar with the type(s) of testing involved. Proper conduct of testing and associated interpretation of results can only be achieved by appropriately trained personnel with access to adequately equipped laboratory facilities. . . . .

Acute Dermal Irritation/Corrosion
(Guideline #404*)

Introductory Information

Prerequisites:

    Solid or liquid test substance
    Chemical identification of test substance

---

*Adopted May 12, 1981. Users of this Test Guideline should consult the Preface, in particular paragraphs 3, 4, 7, and 8.

Purity (impurities) of test substance
Solubility characteristics
pH (where appropriate)
Melting point/boiling point

Standard documents: There are no relevant international standards.

Method

Introduction, purpose, scope, relevance, application and limits of test: In the assessment and evaluation of the toxic characteristics of a substance, determination of the irritant or corrosive effects on skin of mammals is an important initial step. Information derived from this test serves to indicate the existence of possible hazards likely to arise from exposure to the skin.

Definitions: Dermal irritation is the production of reversible inflammatory changes in the skin following the application of a test substance.

Dermal corrosion is the production of irreversible tissue damage in the skin following the application of a test substance.

Principle of the test method: The substance to be tested is applied in a single dose to the skin of several experimental animals, each animal serving as its own control. The degree of irritation is read and scored at specified intervals and is further described to provide complete evaluation of the effects. The duration of the study should be sufficient to evaluate fully the reversibility or irreversibility of the effects observed.

Description of the Test Procedure

Preparations: Approximately 24 hours before the test, fur should be removed by clipping or shaving from the dorsal area of the trunk of the animals. Care should be taken to avoid abrading the skin. Only animals with healthy intact skin should be used.

When testing solids (which may be pulverised if considered necessary), the test substance should be moistened sufficiently with water or, where necessary, a suitable vehicle, to ensure good contact with the skin. When vehicles are used, the influence of the vehicle on irritation of the skin by the test substance should be taken into account. Liquid test substances are generally used undiluted.

Experimental animals: Selection of species. Although several mammalian species may be used, the albino rabbit is recommended as the preferred species.

Number of animals. At least 3 healthy animals should be used. Additional animals may be required to clarify equivocal responses.

Housing and feeding conditions. Animals should be individually housed. The temperature of the experimental animal room should be 22°C (+/- 3°) for rodents, 20°C (+/- 3°) for rabbits, and the relative humidity 30 to 70 per cent. Where the lighting is artificial, the sequence should be 12 hours light, 12 hours dark. Conventional laboratory diets are suitable for feeding and an unrestricted supply of drinking water should be available.

Test conditions: Dose level. A dose of 0.5 ml of liquid or 0.5 g of solid or semisolid is applied to the test site. Separate animals are not required for an untreated control group. Adjacent areas of untreated skin of each animal serve as control for the test.

Observation period. The duration of the observation period should not be fixed rigidly but should be sufficient to evaluate fully the reversibility or irreversibility of the effects observed. It need not normally exceed 14 days after application.

Procedure: The test substance should be applied to a small area (approximately 6 cm^2) of skin and covered with a gauze patch, which is held in place with nonirritating tape. In the case of liquids or some pastes it may be necessary to apply the test substance to the gauze patch and then apply that to the skin. The patch should be loosely held in contact with the skin by means of a suitable semi-occlusive dressing for the duration of the exposure period. However, the use of occlusive dressing may be considered appropriate in some cases. Access by the animal to the patch and resultant ingestion/inhalation of the test substance should be prevented.

Exposure duration is four hours. Longer exposures may be indicated under certain conditions, e.g., expected pattern of human use and exposure. At the end of the exposure period, residual test substance should be removed where practicable, using water or an appropriate solvent, without altering the existing response or the integrity of the epidermis.

Clinical observations and scoring: Animals should be examined for signs of erythema and oedema and the responses scored at 30--60 minutes, and then 24, 48, and 72 hours after patch removal.

Animal Irritancy Testing

Dermal irritation is scored and recorded according to the grades in Table 1. Further observations may be needed, as necessary, to establish reversibility. In addition to the observation of irritation, any serious lesions and other toxic effects should be fully described.

Data and Reporting

Treatment of results: Data may be summarized in tabular form, showing for each individual animal the irritation scores for erythema and oedema at 30–60 minutes, 24, 48, and 72 hours after patch removal, any serious lesions, a description of the degree and nature of irritation, corrosion, or reversibility, and other toxic effects observed.

Evaluation of results: The dermal irritation scores should be evaluated in conjunction with the nature and reversibility or otherwise of the responses observed. The individual scores do not represent an absolute standard for the irritant properties of a material, and they should be viewed as reference values which are only meaningful when supported by a full description and evaluation of the observation(s). The use of an occlusive dressing is a severe test and the results are relevant to very few likely human exposure conditions.

Test report: The test report must include the following information:

- Species/strain used
- Physical nature and, where appropriate, concentration and pH value for the test substance
- Tabulation of irritation response data for each individual animal for each observation time period (e.g., 30–60 minutes, 24, 48, 72 hours after patch removal)
- Description of any serious lesions observed
- Narrative description of the degree and nature of irritation observed
- Description of any toxic effects other than dermal irritation

Interpretation of the results: Extrapolation of the results of dermal irritancy/corrosivity studies in animals to man is valid only to a limited degree. The albino rabbit is more sensitive than man to irritant substances in most cases. The finding of similar results in tests on other animal species may give more weight to extrapolation from animal studies to man.

Literature

(1) WHO Publication: *Environmental Health Criteria 6, Principles and Methods for Evaluating the Toxicity of Chemicals*. Part II (in preparation).

(2) United States National Academy of Sciences, Committee for the Review of NAS Publication 1138, *Principles and Procedures for Evaluating the Toxicity of Household Substances*, Washington, 1977.

(3) Draize, J. H., Woodward, G., and Calvery, H. O. J. Pharmacol. Exp. Ther., 83:377-390, 1944.

(4) Draize, J. H. *The Appraisal of Chemicals in Foods, Drugs, and Cosmetics*, pp. 46-48, Association of Food and Drug Officials of the United States, Austin, Texas, 1959.

(5) Advances in Modern Toxicology, Vol. 4, *Dermato-Toxicology and Pharmacology*, (eds. Marzulli, F. N., and Maibach, H. I.) Hemisphere Publishing Co., Washington-London, 1977.

(6) Draize, J. H. *Appraisal of the Safety of Chemicals in Foods, Drugs, and Cosmetics*:pp. 46-59. Assoc. of Food and Drug Officials of the United States, Topeka, Kansas, 1965.

REFERENCES

1. National Academy of Sciences, Committee for the Revision of NAS Publication 1138. *Principles and Procedures for Evaluating the Toxicity of Household Substances*. National Academy of Sciences, Washington, DC, pp. 23–59 (1977).

2. Code of Federal Regulations. Office of the Federal Registrar, National Archives of Records Service. General Services Administration Title 16, part 1500.40 (1985).

3. Code of Federal Regulations. Title 16, part 1500.41 (1985).

4. Code of Federal Regulations. Title 16, part 1500.42 (1985).

5. McCreesh, A. H., and Steinberg, M. Skin irritation testing in animals. In: *Dermatoxicology*, 2nd ed. (F. N. Marzulli and H. I. Maibach, eds.). McGraw-Hill, Hemisphere Publishing Corp., New York, pp. 147–166 (1983).

6. Steinberg, M. Dermatotoxicology test techniques: an overview. In: *Cutaneous Toxicity*, 2nd ed. (V. A. Drill and P. Lazar, eds.), Raven Press, New York, pp. 41–53 (1984).

7. Omitted in proof.

8. Draize, J. H., Woodard, G., and Calvery, H. O. Methods for the study of irritation and toxicity of substances applied topically to the skin and mucous membranes. *J. Pharmacol. Exp. Ther.* 82:377–390 (1944).

9. Lansdown, A. B. G. An appraisal of methods for detecting primary skin irritants. *J. Soc. Cosmet. Chem.* 23:739–772 (1972).

10. Hood, D. B. Practical and theoretical considerations in evaluating dermal safety. In: *Cutaneous Toxicity* (V. A. Drill and P. Lazar, eds.). Academic Press, New York, pp. 15–30 (1977).

11. Code of Federal Regulations. Title 49, part 173.240 (1985).

12. Organization for Economic Cooperation and Development. *OECD Guidelines for Testing of Chemicals.* Paris, OECD (1981).

13. Roudabush, R. L., Terhaar, C. J., Fassett, D. W., and Dziuba, S. P. Comparative acute effects of some chemicals on the skin of rabbits and guinea pigs. *Toxicol. Appl. Pharmacol.* 7:559–565 (1965).

14. Opdyke, D. The guinea pig immersion test—A 20 year appraisal. *CTFA Cosmetic J.* 3(3):46–47 (1971).

15. Hood, D. B., Neher, R. J., Reinke, R. E., and Zapp, J. A. Experience with the guinea pig in screening primary irritants and sensitizers. *Toxicol. Appl. Pharmacol.* 7:485–486 (1965).

16. Griffith, J. F., and Buehler, E. V. Prediction of skin irritancy and sensitizing potential by testing animals and man. In: *Cutaneous Toxicity* (V. A. Drill and P. Lazar, eds.). Academic Press, New York, pp. 155–173 (1977).

17. Finkelstein, P., Laden, K., and Miechowski, W. Laboratory methods for evaluating skin irritancy. *Toxicol. Appl. Pharmacol.* 7:74–78 (1965).

18. Finkelstein, P., Laden, K., and Miechowski, W. New methods for evaluating cosmetic irritancy. *J. Invest. Dermatol.* 40:11–14 (1963).

19. Motoyoshi, K., Toyoshima, Y., Sato, M., and Yoshimura, M. Comparative studies on the irritancy of oil and synthetic perfumes to the skin of rabbit, guinea pig, rat, miniature swine, and man. *Cosmetics Toiletries* 94:41–48 (1979).

20. Brown, V. K. H. A comparison of predictive irritation tests with surfactants on human and animal skin. *J. Soc. Cosmet. Chem.* 22:411–420 (1971).

21. Uttley, M., and van Abbé, N. J. Primary irritation of the skin: mouse ear test and human patch test procedures. *J. Soc. Cosmet. Chem.* 24:217–227 (1973).

22. Davies, R. E., Harper, K. H., and Kynoch, S. R. Interspecies variation in dermal reactivity. *J. Soc. Cosmet. Chem.* 23:371–381 (1972).

23. Kastner, W. Zur Speziesabhangigkeit der hautvertraglichkeit von kosmetikgrundstoffen. *J. Soc. Cosmet. Chem.* 28:741–754 (1977).

24. Justice, J. D., Travers, J. J., and Vinson, L. J. The correlation between animal tests and human tests in assessing product mildness. *Proc. Sci. Section Toilet Goods Assoc.* 35:12–17 (1961).

25. Marzulli, F. N., and Maibach, H. I. The rabbit as a model for evaluating skin irritants: a comparison of results obtained on animals and man using repeated skin exposures. *Food Cosmet. Toxicol.* 13:533–540 (1975).

26. MacMillan, F. S. K., Rafft, R. R., and Elvers, W. B. A comparison of the skin irritation produced by cosmetic ingredients and formulations in the rabbit, guinea pig, beagle dog to that observed in the human. In: *Animal Models in Dermatology* (H. I. Maibach, ed.). Churchill-Livingstone, Edinburgh, pp. 12–22 (1975).

27. Bartek, M. J., Labudde, J. A., and Maibach, H. I. Skin permeability in vivo: Rat, rabbit, pig, man. *J. Invest. Dermatol.* 56:409 (1971).

28. Campbell, R. L., and Bruce, R. D. Comparative dermatotoxicology I. Direct comparison of rabbit and human primary skin irritation responses to isopropyl myristate. *Toxicol. Appl. Pharmacol.* 59:555–563 (1981).

29. Nixon, G. A., Tyson, C. A., and Wertz, W. C. Interspecies comparisons of skin irritancy. *Toxicol. Appl. Pharmacol.* 31:481–490 (1975).

30. Frosch, P. J., and Kligman, A. M. the chamber scarification test for assessing irritancy of topically applied substances. in: *Cutaneous Toxicity* (V. A. Drill and P. Lazar, eds.). Academic Press, New York (1977).

31. Weil, C. S., and Scala, R. A. Study of intra and interlaboratory variability in the results of rabbit eye and skin irritation tests. *Toxicol. Appl. Pharmacol.* 19:276–360 (1971).
32. Phillips, L., Steinberg, M., Maibach, H. I., and Akers, W. A. A comparison of rabbit and human skin response to certain irritants. *Toxicol. Appl. Pharmacol.* 21:369–382 (1972).
33. Gilman, M. R., Evans, R. A., and DeSalva, S. J. The influence of concentration, exposure duration, and patch occlusivity upon rabbit primary dermal irritation indices. *Drug Chem. Toxicol.* 1(4):391–400 (1978).
34. Ingram, A. J., and Grasso, A. Patch testing in the rabbit using a modified human patch test method. Application of histological and visual assessment. *Br. J. Dermatol.* 92:131–142 (1975).
35. Kligman, A. Quantitative testing of chemical irritants. In: *Evaluation of Therapeutic Agents and Cosmetics* (M. Steinberg et al., eds.). McGraw-Hill, New York, pp. 186–192 (1964).
36. Vinegar, M. B. Regional variation in primary skin irritation and corrosivity potentials in rabbits. *Toxicol. Appl. Pharmacol.* 49:63–69 (1979).
37. Steinberg, M., Akers, W. A., Weeks, M., McCreesh, A. H., and Maibach, H. I. I. A comparison of test techniques based on rabbit and human skin responses to irritants with recommendations regarding the evaluation of mildly or moderately irritating compounds. In: *Animal Models in Dermatology* (H. I. Maibach, ed.). Churchill-Livingston, Edinburgh (1975).
38. Fisher, A. A. *Contact Dermatitis*, 2nd ed. Lea & Febiger, Philadelphia, p. 79 (1973).
39. Holland, B. D., Cox, W. C., and Dehne, E. J. "Prophetic" patch test. Report on results of some 14,000 completed tests performed by the Army Industrial Hygiene Laboratory. *Arch. Dermatol. Syph.* 61:611–618 (1950).
40. Fernstrom, A. I. B. Patch test studies. 1. A new patch test technique. *Acta Dermatol. Venereol.* 34:203–215 (1954).
41. Fernstrom, A. I. B. Patch test studies. 2. Details of the method and practical experience of the pressure test. *Acta Dermatol. Venereol.* 35:420–428 (1955).

42. Anderson, W. A., Shatin, H., and Canizares, O. Influence of varying physical factors on patch test responses. *J. Invest. Dermatol.* 30:77–81 (1958).

43. Lansdown, A. B. G. Animal models for the study of skin irritants. *Curr. Probl. Dermatol.* 7:26–38 (1978).

44. Roper, S. S., and Jones, H. E. An animal model for altering the irritability threshold of normal skin. *Contact Dermatitis* 13:91–97 (1985).

45. Siegel, J. M., and Meltzer, L. Patch tests versus usage tests with special reference to volatile ingredients. *Arch. Dermatol. Syph.* 57:660–663 (1948).

46. Elliott, G. A., and Gray, J. E. Morphologic effects of mildly irritating topical agents. *Toxicol. Appl. Pharmacol.* 16:362–373 (1970).

47. Lanman, B. M., Elvers, W. B., and Howard, C. S. The role of human patch testing in product development programme. *Proc. Joint Conf. Cosmet. Sci., Toilet Goods Assoc.*, Washington, DC, pp. 135–145 (1968).

48. Sullivan, J. B., Strausburg, J. C., and Kapp, R. W. A comparative study of dermal reactions using the intact rabbit skin. *Toxicol. Appl. Pharmacol.* 33:165 (1975).

49. Brown, V. K., and Clarke, R. A. Sulphan blue as an aid to the laboratory assessment of primary skin irritants. *J. Invest. Dermatol.* 45(3):173–176 (1965).

50. Skog, E. Irritant effect of industrial hand cleaners. *Arch. Environ. Health* 7:682–685 (1963).

51. Fisher, L. B., and Maibach, H. I. Effect of some irritants on human epidermal mitosis. *Contact Dermatitis* 1:273–276 (1975).

52. Pitts, E. P., Smerbeck, R. V., and Rieger, M. M. An animal test procedure for the simultaneous assessment of irritancy and efficacy of skin care products. In: *Models in Dermatology*, Vol. 2 (H. I. Maibach and N. Lowe, eds.). Karger, Basel, pp. 209–224 (1985).

53. Van der Valk, P. G. M., Nater, J. P., and Bleumink, E. Skin irritancy of surfactants as assessed by water vapor loss measurements. *J. Invest. Dermatol.* 82:291–293 (1984).

54. Bisgaard, H., and Kristensen, J. K. Quantitation of microcirculatory blood flow changes in human cutaneous tissue

induced by inflammatory mediators. *J. Invest. Dermatol.* 83:184–187 (1984).

55. Tur, E., Tur, M., Maibach, H. I., and Guy, R. H. Basal perfusion of the cutaneous microcirculation: Measurements as a function of anatomic position. *J. Invest. Dermatol.* 81:442–446 (1983).

56. Engelhart, M., and Kristensen, J. K. Evaluation of cutaneous blood flow responses by 133 Xenon washout and a Laser-Doppler Flowmeter. *J. Invest. Dermatol.* 80:12–15 (1983).

57. Staberg, B., Klemp, P., and Serup, J. Patch test responses evaluated by cutaneous blood flow measurements. *Arch. Dermatol.* 120:741–743 (1984).

58. Roper, S. S., and Jones, H. E. A new look at conditioned hyperirritability. *J. Am. Acad. Dermatol.* 7:643–650 (1982).

59. Bjornberg, A. Increased skin reactivity to primary irritants provoked by hand eczema. *Arch. Dermatol. Forsch.* 249:389–400 (1974).

60. Cochran, W. G., and Cox, G. M. *Experimental Designs*, 2nd ed. Wiley, New York, Sections 3.54a, 13.22, and Chapter 11 (1957).

61. Wilcoxon, F., Katti, S. K., and Willcox, R. A. *Critical Values and Probability Levels for the Wilcoxon Rank Sum Test and the Wilcoxon Signed Rank Test*. Florida State University Press, Tallahassee, pp. 1–64 (1963).

62. Snedecor, G. W., and Cochran, W. G. *Statistical Methods*, 6th ed. Iowa State University Press, Ames, pp. 157, 195–196 (1967).

63. Calandra, J. Comments on the guinea pig immersion test. *CFTA Cosmetic J.* 3(3):47 (1971).

64. Opdyke, D. L., and Burnett, C. M. Practical problems in the evaluation of the safety of cosmetics. *Proc. Sci. Section Toilet Goods Assoc.* 44:3–4 (1965).

65. Klauder, J. V. Patch test study to determine cutaneous reaction to new compounds. *Arch. Environ. Health* 1:43–52 (1960).

66. Guillot, J. P., Martini, M. C., and Giauffret, J. Y. Safety evaluation of cosmetic raw materials. *J. Soc. Cosmet. Chem.* 28:377–393 (1977).

67. Kligman, A. M., and Wooding, W. M. A method for the measurement and evaluation of irritants on human skin. *J. Invest. Dermatol.* 49:78–94 (1967).

68. Idson, G. Topical toxicity and testing. *J. Pharm. Sci.* 57(1):1–11 (1968).

69. Schwartz, L., and Peck, S. The patch test in contact dermatitis. *Public Health Rep.* 59:546 (1944).

70. Mills, O. H., Swinyer, J., and Kligman, A. M. Assessment of irritancy of skin cleansers. Scientific Exhibit presented at the XLIII Annual meeting of the American Academy of Dermatology, Washington, DC, December 1984.

3
"Predictive" Animal Sensitization Assays: What Do They Mean?

KLAUS E. ANDERSEN *Department of Dermatology, University Hospital, Odense, Denmark*

HOWARD I. MAIBACH *University of California Hospital, San Francisco, California*

WHY PERFORM SENSITIZATION ASSAYS?

Allergenicity evaluation is needed to reduce the hazard of occupational and consumer exposure from chemicals and products that may contact the skin.

The magnitude of the contact allergy problem is seen in epidemiological investigations, which have shown different figures due to variations in methodology. Skin diseases made up about half of the reported industrial diseases in the United States, and contact dermatitis was the most prevalent diagnosis, accounting for 32% of dermatological occupational diseases in New York (1).

This chapter was adapted from: *Current Problems in Dermatology*, Vol. 14, pp. 263-290 (Karger, Basel, 1985).

In Denmark 7% of patients seeking outpatient dermatological care had contact dermatitis (2). Topical drugs and cosmetics were frequent causes of allergic contact dermatitis (3,4). However, dermatological studies underestimated the frequency of adverse reactions from cosmetics, because most problems are solved by consumers themselves by discontinuing use of a product or by other trial-and-error methods. The consumer experiences more reactions than are documented in government or other reports (5,6).

Most of our epidemiological data has been gathered in group studies, such as from the International Contact Dermatitis Research Group (ICDRG) and the North American Contact Dermatitis Research Group (NACDRG) (5,7). These data have only infrequently been corrected for technical artifacts such as the excited skin syndrome (ESS) (8,9). Normal population studies have been performed even less frequently (10,11).

A second reason to perform allergy tests is because manufacturers can and are often held responsible for side effects produced by products not properly tested.

Third, the demand to avoid chemicals and products that carry a health hazard is increasing from consumer organizations, unions, and governmental agencies. The appearance of official recommendations regarding guinea pig allergy tests and textbooks in dermatotoxicology is a result of this demand, e.g., OECD Guidelines for Testing of Chemicals (1981) and recent monographs on guinea pig allergy tests (12–14).

The requirement for allergenicity tests is extended from chemicals with industrial use to consumer products with various fields of applications—toilet paper, food wrapping—which may indicate a hazard if allergenic ingredients in sufficient concentrations were present. The development of guinea pig sensitization assays with special suitability for testing of finished products is a result.

GUINEA PIG SENSITIZATION ASSAYS

All guinea pig assays use the same approach: an induction phase, followed by a rest period of about 2 weeks, and subsequently a challenge test to demonstrate whether or not sensitization occurred. Predictive contact allergy tests can be performed in laboratory animals as well as in humans. The preferred animal species is the albino guinea pig. Mouse models have been designed, but their predictive value has not yet been established (15–17).

Male or female (except pregnant) outbred albino guinea pigs are traditionally used. Females may have the advantage of being less aggressive than males (18). Animals 1–3 months of age weighing about 350–400 g are preferred. Considering the genetic influence on susceptibility to sensitization (19–21), sensitivity of the animal strain to be used in sensitization studies should be verified with moderate, well-known sensitizers. As with other biological tests, standardization of maintenance and handling procedures should be ascertained to assure reproducibility of test results (18,22).

Most test methods prescribe the use of 20–40 healthy young adult albino guinea pigs. They are divided into test groups and a control group. The simultaneous use of sham-treated animals is highly recommended in order to have proper controls and the possibility of blind-reading of challenge responses.

Figure 1 shows the time schedules and method for induction and challenge for each assay.

The challenge reactions are usually graded 2–3 times, with 24-hr intervals, according to a four- to seven-point ranking scale giving the degree of erythema and edema, i.e., Magnusson and Kligman's four-point scale (18) (Table 1). Nonspecific effects are greater shortly after removal of the patch. As in human patch testing, much information is gained by delayed readings.

The test methods can be grouped according to different criteria: the use of Freund's complete adjuvant and the application route for induction (Table 2). The latter is the basis for the order of the following brief description of the methods.

Intradermal Methods

The Draize test (23) is a variant of the classical Landsteiner technique (24). Induction is achieved by 10 intradermal injections of the test material in a concentration of 0.1% in a suitable vehicle. Challenge is also done by intradermal injection of the test material preparation.

The Draize test is easy to perform and useful as a screening method for detecting "strong" sensitizers, while many "weak" sensitizers remain undetected by this method (18). The use of an induction concentration fixed at 0.1% is unfounded. The test is applicable to chemicals, not to finished products. The test has been modified several times to improve the sensitivity. Voss (25) increased the concentrations for induction and challenge. Prince and Prince (26) replaced the intradermal injections with topical applications without increasing the sensitivity of the method.

Sharp (27) examined 69 perfume ingredients using a modification where he increased the induction exposure and shortened

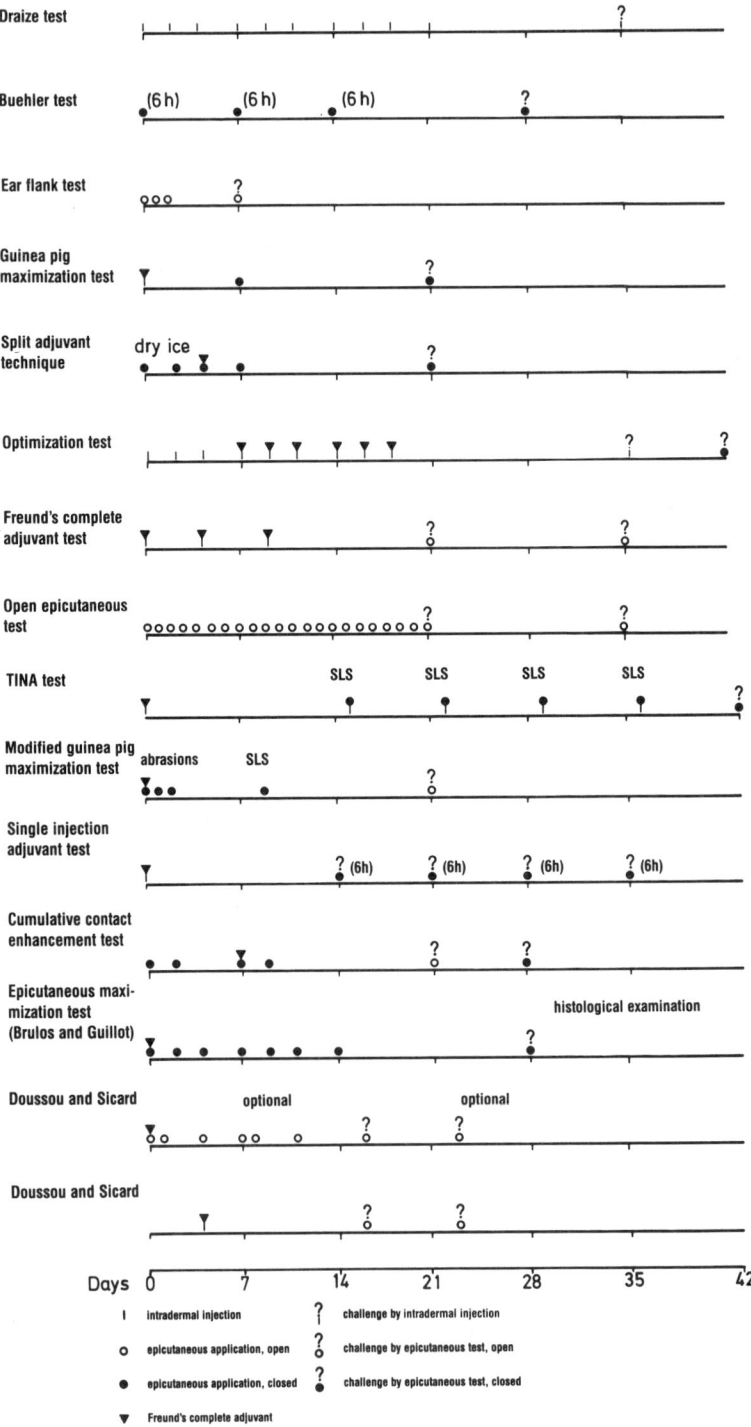

Table 1 Magnusson and Kligman's Grading Scale for the Evaluation of Challenge Patch Test Reactions

---

0 = no visible change

1 = discrete or patchy erythema

2 = moderate and confluent erythema

3 = intense erythema and swelling

Table 2 Guinea Pig Allergy Tests Grouped According to the Application Route for Induction, the Use of Freund's Complete Adjuvant (FCA), and Suitability for Testing of Finished Products

	Uses FCA	Applicable for testing finished products
Intradermal methods		
Draize test	−	
Optimization test	+	
Freund's complete adjuvant test	+	
Single injection adjuvant test	+	
Epicutaneous methods		
Buehler test	−	Yes
Ear flank test	−	Yes
Open epicutaneous test	−	Yes
Modified guinea pig maximization test	+	Yes
Cumulative contact enhancement test	+	Yes
Epicutaneous maximization test	+	Yes
Split adjuvant technique	+	Yes
Methods using both application routes		
Guinea pig maximization test	+	
TINA test	+	
Dossou and Sicard test	+	Yes

Figure 1 Guinea pig contact allergy tests. Schedule for induction and challenge.

the test period. Instead of 10 induction injections using the challenge concentration, he gave the same total dose in 1 day, divided into four intradermal injections at 2.5 times the challenge concentration. If the first challenge 2 weeks later was negative, the procedure was repeated. This improved the sensitivity, but in spite of the modifications, the Draize test remains less sensitive than the methods using Freund's complete adjuvant.

The optimization test (14,27) is similar to the Draize test but differs by using a Freund's complete adjuvant emulsion as vehicle for the intradermal injections in the second and third weeks. The concentration for induction is fixed at 0.1%. Repeated challenges are used: First, intradermal challenge is made, with a quantitative reading of reaction diameter and thickness as a commendable feature. Second, epidermal occlusive challenge 2 weeks later is included to confirm the initial results and get more information about the test substance. The optimization test appears to be more efficient than the Draize test in detecting sensitizers, and good agreement in test results has been obtained with the guinea pig maximization test.

Freund's complete adjuvant test (FCAT) and the open epicutaneous test (OET) (29) are used as supplementary test methods for the evaluation of allergenicity. The FCAT is a sensitive variant of the intradermal test methods. The test substances are emulsified in FCA to a suitable concentration (5–50%) and injected intradermally 3 times in 10 days.

Challenge is performed by open application of the test substance in the minimal irritating concentration and dilutions thereof in order to obtain a dose-response relationship for the elicitation phase. The challenge reactions are evaluated by an all-or-none criterion. Like the Draize test, it is not applicable to finished products. The FCAT is technically simple, and like the GPMT, the FCAT rates as sensitizers a number of compounds that were judged as very weak sensitizers from human predictive and diagnostic patch test experience.

The single injection adjuvant test (SIAT) (30) was slightly less sensitive than the guinea pig maximization test in a comparative study using 19 allergens. The induction procedure is condensed to the single injection of the test material in Freund's complete adjuvant. The occluded challenge patch tests are left on the skin for 6 hr, which might decrease the strength of the reactions. Repeated challenges with 1-week interval is suggested. The method is suitable for testing of chemicals and not for finished products.

Epicutaneous Methods

The closed patch test of Buehler (31) is a tempting guinea pig model for allergenicity evaluation, because it mimics the consumer's exposure condition: repeated topical applications, finished products can be tested, and the induction concentration is increased to the limit where it does not provoke excessive irritation. The method was empirically designed to mimic the human repeated insult patch test (32,33).

However, how sensitive is the Buehler test? A heavy metal sensitizer such as nickel was not detected by this test. In a comparison of five guinea pig methods, Marzulli and Maguire (34) found the Buehler test to be the least sensitive, when they studied a number of possible cosmetic allergens that have been tested on humans. In contrast to the original description, they occluded the patch tests with a circular elastic bandage, while Buehler restrained the animals and occluded with a rubber dam. This could influence the quality of the occlusion (35).

Experience with the ear flank test was reported by Stevens (36), who had tested more than 100 chemicals. The test procedure is rapid and easy. Open applications to the ears is used for induction, and clipping of the flank is done only once prior to the challenge on day 7. No further experience with this test has been published.

The open epicutaneous test (OET) (29) uses for induction repeated uncovered topical applications. Freund's complete adjuvant is not applied, which makes the test less sensitive than those applying this immune enhancer. The sensitivity of the test was comparable to that of the guinea pig maximization test for 32 fragrance materials (37), but in other experiments it was less sensitive (38,39).

An advantage of the OET is the prescribed use of progressive dilutions of the test compound in separated groups of animals. This procedure determines a dose-response relationship both for the induction phase and, by the use of different challenge concentrations, for the elicitation phase.

The modified guinea pig maximization test of Sato et al. (40) was developed to test cosmetic ingredients, which were difficult to dissolve or suspend appropriately for the intradermal injection. The skin at the induction site is abraded before the topical closed test material application. Freund's complete adjuvant is used. The method seems to be sensitive.

The cumulative contact enhancement test (CCET) (41) is designed to test finished household products, as well as their ingredients. It combines the use of Freund's complete adjuvant with covered epidermal applications. In contrast to the modified guinea pig maximization test (40), the epidermal barrier is not abraded. Tsuchiya and co-workers found the highest response rate when the adjuvant was injected before the third induction patch test, and not at the start of the test. The method appears to be sensitive, as are the other assays using Freund's complete adjuvant.

The epicutaneous maximization test (EMT) (42) is a modification of the method of Brulos et al. (43). It is applicable for cosmetics, and again, it combines the use of Freund's complete adjuvant with topical closed applications. When the method of Guillot/Brulot was compared to the guinea pig maximization test, the split adjuvant technique, the optimization test, the open epicutaneous test, and the method of Doussou and Sicard using six different allergens, the methods were qualitatively similar in their capacity to identify allergens (42).

In the design of the EMT, a special feature is made of histological studies of challenge reactions. Biopsies for microscopical examinations are taken from 49 to 96 hr after application of the challenge patches. The use of histology to distinguish between allergic and irritant reactions presupposes expert knowledge, which makes it an unrealistic approach in most laboratories dealing with dermatotoxicology. Recent studies on the histological and cellular response in allergic and irritant contact dermatitis in humans could not distinguish between allergic and irritant cell reactions (44,45). This is in agreement with the evidence that only about 10% of the infiltrating cells in a contact sensitivity skin reaction are, in fact, sensitizied specifically. The remaining cells are attracted to the site nonspecifically (46).

Maguire's split adjuvant technique (47,48) was designed to test preparations intended for topical application in humans, but any chemical can be tested with his method. The allergenicity rating of a preparation was made relative to the allergenicity of an alternative product. The test utilized Freund's complete adjuvant injections and four occlusive topical applications of the test material. The technique appears to be sensitive, but time consuming. For grading of test reactions a seven-point ranking scale requires considerable experience and judgment and may be an exaggeration for routine use, as reading variations both between and within observers could be significant. Several variations of the method have been published (25,38).

Methods Using Both Application Routes

The guinea pig maximization test (GPMT) developed by Magnusson and Kligman (18) is considered among the most sensitive procedures for detecting the potential sensitizing capacity of a compound. The procedure was developed through a systematic series of experiments, though which the authors "maximized" each single component of the technique to achieve maximum level of sensitization. This logical background for the procedure is one factor that helped make the test so widely used.

The GPMT combines the use of intradermal injections, Freund's complete adjuvant, and topical applications under occlusion. The test was designed to be as simple as possible, to reduce the need for manpower; therefore, only two induction treatments are prescribed (Fig. 2).

The test concentrations for induction should be the highest ones to produce mild to moderate irritation. If the test material is not locally irritating, the skin over the induction area is treated with sodium lauryl sulfate 10% in petrolatum 24 hr before the topical induction on day 7.

The test gives the potential sensitizing capacity of a chemical. In many instances, substances that are potent sensitizers in this assay can be used without giving rise to an epidemic of contact sensitivities, because the conditions for use of the substances (concentration, mode, and route of exposure) are much less hazardous, when related to the test situation (39,49).

Finished products composed of several ingredients should not be tested with this method. Usually one or two ingredients are potential allergens, while the rest are water and nonallergenic substances, which influence the choice of induction concentration. A locally irritating, nonallergenic ingredient will decrease the test concentration for an allergenic ingredient that is mildly irritating, if they are tested together. The decreased test concentration might prevent detection of the allergen.

The TINA test (Tierexperimenteller Nachweistest) (50) was developed from the method described by Polak and Turk (51). The procedure is time consuming, sensitive, and combines intradermal and topical administration of the test substance and the use of Freund's complete adjuvant. The procedure is applicable for all compounds that can be made soluble in a suitable vehicle. Whether the technique has advantages over one of the more simple and apparently equally sensitive test methods remains to be established.

The method of Doussou and Sicard (52) is designed for testing cosmetics. They utilize two parallel groups of test animals: one

induction

Day 0

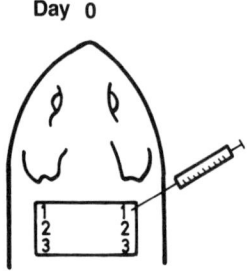

intradermal injections

1 : 0.1 ml FCA
2 : 0.1 ml test material
3 : 0.1 ml test material/FCA

7

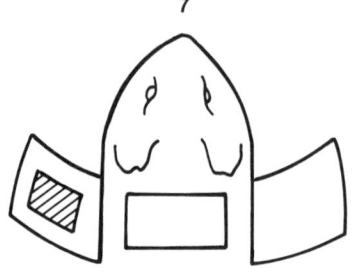

48 hours closed patch test

challenge

Day 21 and 35

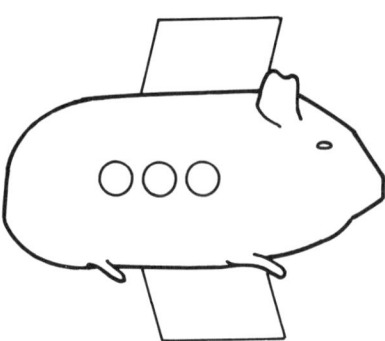

24 hours closed patch test

Figure 2 The guinea pig maximization test.

treated with three open topical applications, the other treated with intradermal injection on day 5 into the footpad of a hind leg. Both groups receive Freund's complete adjuvant in the hind leg. These investigators allow for modifications in the induction procedure according to the intended use of the product tested.

FACTORS AFFECTING INDUCTION AND ELICITATION OF CONTACT ALLERGY

To act like sensitizers, low-molecular-weight chemicals called haptens must first penetrate the skin and bind irreversibly to large carrier molecules, i.e., skin protein (46,53,54). The hapten-protein complexes are then recognized as antigens by the mononuclear phagocytic cells present in the epidermis and dermis. The Langerhans cells in epidermis and dermis constitute a prominent part of the antigen recognition system (55,56).

Both for the consumer and in the experimental situation, the considerable variation in exposure conditions and individual susceptibility is important for the development of contact sensitivity.

Various factors related to the allergen, the individual, and the environment are important for the development of contact allergy (Table 3). Some allergen-related and a few subject-and-environment-related factors are discussed here.

Table 3 Factors Affecting Induction and Elicitation of Contact Allergy

Allergen
  Physicochemical properties
  Dosage and concentration
  Vehicle
  Mode and route of exposure

Human
  Preexisting eczema
  Genetic predisposition
  Sex and age

Environment
  Temperature
  Humidity
  Hygiene
  Season of the year

Physicochemical Properties

Contact allergens are low-molecular-weight compounds, i.e., with a molecular weight under about 1000 daltons.

Landsteiner and Jacobs (57) initially found that only those chloro- and nitro-substituted benzene derivatives that were capable of binding covalently to aniline could sensitize guinea pigs. The molecular basis behind the reactions between haptens and skin protein is discussed in detail in Dupuis and Benezra's monograph (54).

The "irreversible binding theory" is an oversimplification, as most allergenic chemicals do not react, or cannot be demonstrated in vitro to react, covalently with proteins. In such cases other possibilities may explain the allergenicity. Metabolites, degradation products, or impurities present in the substances may be the chemically reactive agents responsible for the immunogenicity.

It is currently impossible to predict on the basis of structure alone whether a simple chemical is capable of being a sensitizer, but high chemical reactivity makes a chemical a suspect allergen. However, other physicochemical factors, namely, irritancy potential, lipophilicity, solubility, and effective dose remaining at the induction skin site, influence antigenicity (58,59).

Gold salts are strong allergens experimentally, but since gold is not solubilized by sweat, contact dermatitis from it is rare (60). Trivalent chromate is far more widespread in the environment but rarely causes sensitization and rarely elicits dermatitis in persons sensitized by hexavalent chromate (61,62).

Benezra and co-workers (63) are systematically coding the world literature on allergic contact dermatitis in guinea pig and humans in an attempt to more precisely define structure-activity relations (SAR). The publication should be consulted for experimental detail. Basically, it is hoped that the power of the computer can be used to supplement current theories as to what humans and animals recognize as antigens or nonantigens. The paper presents references documenting use of the method in chemical carinogenicity SAR studies.

Dosage and Concentration

Within limits, it is established that higher concentrations and higher dosages increase the risk of sensitization (18,64,65). The incidence of paraben sensitivity among eczema patients was about 1% in Denmark 20 years ago, probably because of frequent use of topical antifungal agents containing up to 5% paraben (Amycen) (66). Paraben sensitivity is now rare and occurs in risk-group patients with leg ulcers, stasis dermatitis, and skin disorders (67).

For allergens showing a strong sensitizing potential experimentally in guinea pigs, there may be a threshold concentration below which the allergens can be used in consumer products with impunity. Some thiazole preservatives, e.g., Kathon CG are widely used in low concentrations in spite of being potent guinea pig sensitizers (68). Recently, reports documenting frequent contact allergy to this preservative have appeared from some countries, but it is still rare in other countries (69,70). Whether the situation will change with continued usage remains to be seen. Chromate sensitization in construction workers appears after a considerable delay (71—73).

When sensitization has developed, there is a threshold for elicitation. In formalin-sensitized individuals, Jordan et al. (74) applied formalin-containing material to the axilla: 1/3 of their subjects reacted to a 72-hr patch test with a 100 ppm formaldehyde solution, corresponding to 2 µg formaldehyde per patch.

Repeated exposure to allergens may even lead to immunological unresponsiveness (tolerance), as demonstrated in several animal experiments using picryl chloride and dinitrochlorobenzene as allergens (46,75,76).

Specific unresponsiveness can be induced by oral, intravenous, and topical administration of the allergen prior to the sensitization treatment; as well as by feeding or topical application of a structurally related but nonsensitizing chemical. The phenomenon is partly explained by induction of specifically activated cyclophosphamide-sensitive suppressor cells.

Nonallergic humans treated with repeated intramuscular injections of oil derived from poison oak were completely tolerant to a sensitizing dose applied on the skin. After 1 year the degree of tolerance decreased, but the subjects remained clinically hyporeactive for several years (77). How this phenomenon affects clinical contact allergy is unknown.

Vehicle

The bioavailability of a chemical is influenced by the choice of vehicle. Vehicle effects on diagnostic and predictive patch tests are well documented; e.g., formalin is tested in water, most other standard allergens in petrolatum (78—80). Based on the consumption of different neomycin preparations and the occurrence of sensitivity, Hjorth and Thomsen (81) suggested that neomycin was more allergenic in ointments than in creams, powders, and lotions.

A new type of epicutaneous test—the TRUE test—is under clinical evaluation (82). The allergens from the standard series are incorporated in a gel which is applied as a film on a polyester membrane and packed for later use. This new test method seems to have several advantages: The amount of allergen is standardized, and the stability and durability of the allergens are checked and declared. The bioavailability of the allergens is higher than with the current test substances incorporated in petrolatum.

The vehicle, together with such factors as temperature and pH, affects the release of allergen. The amount of formaldehyde set free from a group of industrial formaldehyde releasers was higher when they were mixed with water than with petrolatum (79). Temperature and pH influenced the release of formaldehyde from an acrylate tape and the cosmetic preservative germall 115 (imidazolidinyl urea) (83,84).

In guinea pig allergy tests there is no vehicle that is optimal for all substances. Most investigators prefer to dissolve the test substance because dispersions are prone to form a sediment, making exact dosing difficult. Magnusson and Kligman (18) found 70% ethanol superior to petrolatum, propylene glycol, and acetone in sensitization to DNCB. For intradermal sensitization, ethanol was unsatisfactory and propylene glycol or a vegetable oil was preferable.

Contradictory results regarding the sensitizing potential of (meth)acrylates may partly be explained by differences in the choice of vehicle (85,86).

However, no simple relationship appears between the induction of contact allergy and the bioavailability of an allergen. In cumulative contact enhancement tests with chlorocresol 5% in four different vehicles, the greatest percutaneous absorption appeared from an aqueous suspension that gave the highest sensitization rate, while an olive oil/acetone preparation showed a proportionally low chlorocresol bioavailability, in spite of the fact that it produced an equivalent sensitization rate (87).

Mode and Route of Exposure

Certain anatomical locations are more susceptible to the development of allergic contact dermatitis; e.g., allergic textile resin dermatitis occurs at sites of sweat and friction (88). When elicited by a shirt it starts at the borders of the axillae and parts of the upper trunk. Shampoo dermatitis often appears on the hands and neck before it is visible on the scalp. Glutaraldehyde 25% in water was tolerated on the soles of six subjects, while a 2.5% solution applied to the antecubital fossa gave a positive provocative use test (89). These differences in sensitized subjects are only partially related to the variation in regional human skin penetration. Scrotal skin is about 400 times more penetrable to radiolabeled hydrocortisone than the foot sole (90).

Occlusion increases skin penetration of 14C-hydrocortisone about 10-fold compared to an open application of the same dose (91). The enhancing effect of occlusion occurs in contact allergy, diagnostically as well as experimentally (18). Therapeutically, occlusion increases drug effects (topical effects of corticosteroids) and also the risk of side effects. Patients with leg ulcers, stasis dermatitis, and perianal skin disorders are more prone to develop contact allergy to topical drugs (67). This is explained by the frequent applications and bandaging or skin-to-skin contact. However, it has become obvious recently that in some instances occlusion does not enhance penetration (Maibach, unpublished data).

Allergen-containing products that are left on the skin are more hazardous than rinse-off products, because the short contact time may reduce penetration of the allergen below the limit necessary for induction or elicitation of sensitivity. In experimental studies in guinea pigs excision within hours of the site of allergen application decreased the frequency and the degree of sensitivity (92, 93). The time course relationship will, of course, vary from allergen to allergen, depending on a number of factors, e.g., vehicle release, absorption kinetics, cellular distribution, substantivity, volatility, resistance, and binding (94).

Preexisting Eczema

Leg ulcer patients are disposed to contact allergy not only due to frequent drug applications and bandaging, but also because of the presence of diseased skin with a reduced barrier function (7). In an epidemiological study, Menné et al. (11) showed that hand eczema patients has an increased risk of developing nickel allergy.

Genetics

Contact sensitivity in laboratory animals is genetically controlled. In the guinea pig, for example, it was found that one strain responded well to $K_2Cr_2O_7$ and BeF_2 but not to $HgCl_2$ while another strain responded poorly to $K_2Cr_2O_7$ and BeF_2 and well to $HgCl_2$ (20). The major histocompatibility complex (MHC) plays an important role in the expression of delayed hypersensitivity in the mouse (95).

By selective breeding of guinea pigs, Chase (19,20) established animal strains refractory or less susceptible to sensitization with strong allergens than randomly bred animals. Stampf et al. (96) found the Himalayan strain of guinea pigs resistant to sensitization with alantolactone and isoalantolactone when they used the open epicutaneous test method. The Hartley and Pirbright strain pigs became sensitized.

In the guinea pig maximization test formaldehyde showed a stronger potential in Stockholm than in Copenhagen, probably owing to differences between the guinea pig strains (97).

A study of nickel allergy in the female twin population showed that the heritability of nickel allergy was about 60%, indicating the proportion of the total phenotypic variance (genetic and nongenetic) which was caused by additive genetic variance (98).

Walker et al. (99) demonstrated that children of patients sensitized to 2,4-dinitrochlorobenzene (DNCB) and p-nitrosodimethylaniline (NDMA) became sensitized at a higher rate than did children whose parents were not sensitized.

Environment

Menné and Holm (100) also studied the influence of environmental factors in female twins with hand eczema and nickel allergy. They concluded that exposure was the most important factor for the association between nickel allergy and hand eczema.

In his review on the climate of contact dermatitis (101), Calnan emphasized the influence of temperature and humidity on irritant as well as allergic contact dermatitis. The environmental conditions of work in a factory are important. With proper hygiene, instruction, and information of workers, protective devices, and possibly changed procedures, the hazards of working with allergenic substances can be reduced considerably. The seasonal influence is evident for contact allergy to plants, i.e., *Primula obconica* (102).

WHICH TEST SHOULD BE USED?

There is no simple answer. There have not been enough studies comparing a wider range of chemicals to establish the efficiency of the different test methods in classifying contact sensitization potential and the hazard for humans. Most authors found their own method comparable to other test methods, when they examined various ranges of chemicals (30,34,37,38,42,103–106).

Table 4 shows selected comparative test results.

A sufficient variety of test methods are available. Instead of attempts to modify the techniques, one should concentrate on standardization of the performance of the current tests in order to make their results more comparable. Any unknown chemical or product can be tested for its properties by one or a few of the methods presented.

In general terms, the methods applying Freund's complete adjuvant are sensitive compared to the nonadjuvant methods. In some cases, substances that are potent sensitizers in adjuvant tests show little human sensitization in spite of extensive use in the environment (39,49,79). On the other hand, the nonadjuvant methods may not detect several important contact allergens (28, 34). In selected cases, therefore, screening tests with guinea pigs using both adjuvant and nonadjuvant techniques are justified to get a better characterization of the hazard.

Regarding the choice of test design, the potential routes of human exposure, as well as the magnitude and frequency of exposure, have to be approximated. Is the substance meant for repeated human exposure or is it anticipated to be contained in a closed system, escaping only by accident? The physicochemical characteristics of the product influence the choice of tests and the vehicle.

Some guinea pig allergy tests are designed for the evaluation of chemicals and product components and are not suitable for testing of finished products, e.g., the guinea pig maximization test, the optimization test; while others are designed for testing of finished products as cosmetics and topical drugs, e.g., the split adjuvant technique, the modified guinea pig maximization test, the cumulative contact enhancement test, and the epicutaneous maximization test.

CHOICE OF INDUCTION CONCENTRATION

This question requires additional comments. Some test methods prescribe the use of fixed induction concentrations, e.g., 0.1% in the Draize test, and the optimization test, while others advocate the use of a moderately toxic or locally irritating concentration, e.g., the guinea pig maximization test. The concentration chosen should not interfere with the health of the animals.

Table 4 Comparative Results of Sensitization Assays in Guinea Pigs[a]

	Draize test	Optimization test	Freund's complete adjuvant test	Single injection adjuvant test
Goodwin et al. (1930)				
2,4-Dinichlorobenzene	100[b]			100
Formaldehyde	10[b]			100
2-Mercaptobenzothiazole	0[b]			0
Potassium dichromate	0[b]			50
Nickel sulfate	0[b]			20
Benzocaine	0[b]			0
Chlorhexidine gluconate	0[b]			10
Ethyl paraben	0[b]			10
Guillot et al. (1942)				
Dihydrocoumarin		100	100	
p-Phenylenediamine		5—65		
Formaldehyde		10—100	30	
Penicillin G		33—100	67—100	
Benzocaine		0—90	40—50	
Propylene glycol		0	0	
Marzulli and Maguire (1934)				
Dimethylsulfoxide	0			
Hydroxycittronellal	0			
Formaldehyde	33			
Dowicil 200	0			
Germall 115	0			

For details regarding test technique and choice of concentration, see the references.
[a]The proportion of guinea pigs sensitized is shown (values are in percent.
[b]Modified Draize test.

Another approach is to increase the concentration of the alleged allergen used for induction over that expected to occur during human use or misuse. Cosmetics are generally composed of many ingredients, most of which have a history of safe usage,

Buehler test	Open epicutaneous test	Epicutaneous maximization test	Split adjuvant test	Guinea pig maximization test	Doussou and Sicard test
				100	
				100	
				60	
				100	
				10	
				0	
				20	
				0	
	88	100	100	100	33–75
			100	80–100	
	0–38	35–60	0–70	100	0–50
	0–50	50–95	100–80	90–100	0–17
	38–63	10–45	40–60	60–85	0
	0	0	0	0	0
0			0	0	
0			17	27	
0			7	18	
0			17	55	
0			0	10	

and one or two components that may be allergenic. In the test situation the level of these possible allergenic components can be increased 10-fold to 100-fold and the remaining ingredients can be kept at their in-product level (28,107). If a minimum sensitizing concentration of the allergen is determined, a so-called safety factor can be used to establish the use concentration of the allergen in a finished product. However, the size of the safety factor is arguable.

In the open epicutaneous test, Klecak et al. (29) include several concentrations that result in a dose-response relationship giving more insight into the sensitizing capacity of the test substance. The relationship between the induction doses and the sensitization frequencies seems to be nonlinear (nonmonotonous), suggesting that a chemical may have an optimal induction concentration above or below which a decreased sensitization rate is obtained.

In a study of formaldehyde sensitivity using the guinea pig maximization test in two animal strains we found a nonlinear dose-response relationship (Fig. 3). The intradermal induction concentration was crucial for the sensitization rate, while the topical induction showed no significant effect. The maximum sensitization rate calculated from the fitted curves was 80% in Copenhagen and 84% in Stockholm following intradermal induction with 0.65% and 0.34% formaldehyde, respectively (97).

The nonlinear dose-response relationship found for formaldehyde is concordant with previously reported data for picryl chloride and dinitrochlorobenzene (92), sultones (59), p-nitrobenzyl compounds (108), and to some extent for chlorocresol (39). Stadler and Karol (109) found a similar nonlinear dose-response relationship for dicyclohexylmethane-4,4'-disocyanate and picryl chloride in BALB/cBy mice, while the dose-response relationship was linear for the same allergens in groups of English smooth-haired guinea pigs. However, on a body weight basis, the mice were treated with higher doses of allergens than the guinea pigs. The animals were sensitized by a single open application, and challenge was performed once at day 5 for the mice and at day 7 for the guinea pigs. Repeated challenges might have detected a "down-regulation" effect in some of the exposure groups as seen with chlorocresol in guinea pig maximization tests (110). The level of contact sensitivity in guinea pigs is determined by a balance between activated effector and suppressor cells, and the balance is influenced by the dose given (92); however, decreased sensitivity in the highest exposure groups may also be related to systemic toxicity if the animals appear sick or show weight loss during the test period.

The nonlinear relationship between induction concentration and sensitization response is more pronounced in the methods utilizing Freund's complete adjuvant. When a chemical is injected onto the guinea pig together with adjuvant, the qualitative aspects of the

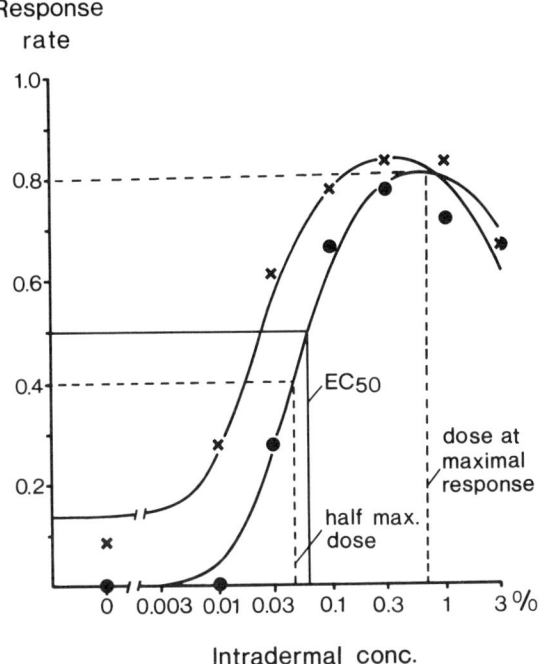

Figure 3 Observed response rates and best-fitting logistic curves for formaldehyde sensitivity (72-hr reading) in the guinea pig maximization test in two laboratories (●---●, Copenhagen; x---x, Stockholm). The maximal response rate and the corresponding formaldehyde concentration, the concentration at the half-maximal response, and the EC_{50} for the Copenhagen results are shown. For further details see Ref. 97.

immune response may be changed, compared to the situation following epicutaneous application. The animals may show an altered production of antibodies and/or a changed balance between the different subsets of lymphocytes (e.g., effector and suppressor cells) (92,111). The results obtained with the help of adjuvants may therefore to some extent exaggerate the allergenicity risk connected with materials used only epicutaneously.

It might improve the value of guinea pig allergy tests if dose-response determinations were employed as a routine procedure and it would be easier to compare results from different investigators and laboratories. The choice of one single concentration is fortuitous in relation to the concentration giving maximal sensitization. It may not be necessary to enlarge the number of animals required for each test material. Instead of using 20 test animals and 20 sham-treated controls, one could use 40 guinea pigs divided into a control group and four to seven test groups, each treated with different induction concentrations. The decrease in group size is counteracted by the increase in the number of groups. Valid calculations are still possible (97).

OTHER SOURCES OF RESULT VARIATION

Guinea pig allergy tests have several sources of errors or causes of variation between the results from different laboratories. Besides the choice of induction concentration, which has been discussed separately, the following factors are important: ambient test conditions, quality of occlusion, reading, and the training of the investigator.

The guinea pig responsiveness vary according to genetic factors, age, diet, season of the year, health of the animals, and stress (21). Recently, Maisey and Miller (16) found that vitamin A acetate augmented the sensitivity of BALB/C mice to environmental allergens, probably due to an increase in the number of antigen-presenting cells in lymph nodes.

The seasonal influence seems to be minimal with the strong contact sensitizers, e.g., dinitrochlorobenzene, and more obvious with the weaker contact sensitizers. Modern animal houses supplying constant temperature, humidity, and light cycle may minimize the seasonal variation.

The importance of proper occlusion is well known from both diagnostic and predictive patch tests (18,112). Ritz and Buehler (107) restrain their animals for 6 hr to assure proper occlusion with a rubber dam. They claim that an elastic bandage circulating the trunk is ineffective. For the challenge procedure the quality of occlusion can be checked if the Finn Chamber (Epitest Ltd., Helsinki, Finland) or a similar aluminum cup is used as patch test unit. Skin indentations from the cup flange immediately after removal of the patches assured proper occlusion.

Reading of the skin reactions is usually accomplished by scoring according to an arbitrary reading scale. This gives the ratings a subjective element. Simultaneously, sham-treated control animals should always be included in guinea pig allergy tests to allow blind reading of the challenge response. Interindividual differences between raters are expected (113,114). The use of sham-treated control animals and blind reading counteracted a possible "excited skin syndrome," which might explain false-positive challenge reactions (8,9,115).

Nonspecific effects are greatest immediately after removing the patch. Much information is gained by delayed readings, as in the case of human patch testing.

The uncertainty of readings has to be added to the variations in technique and animals. This can only be improved by the training and experience of the staff. Quantitation of guinea pig contact allergy by measuring changes in skin blood flow and skinfold thickness may be a useful supplement to visual scoring as a measure of interobserver and interlaboratory differences (116).

The staff planning and conducting guinea pig allergy tests should be adequately trained because the data are only as reliable as the people who have generated them. Consideration should be given to whether or not a protocol can be carried out properly with the staff available. The guidelines published by the Food and Drug Administration (117) for the conduct of nonclinical laboratory studies are an attempt to improve the overall quality of laboratory work.

MOUSE SENSITIZATION ASSAYS

Mice have for decades been the preferred animal species among research immunologists for the study of delayed-type hypersensitivity (DTH). They are cheap, easy to handle, and available in several inbred strains. Further, monoclonal antibodies directed against the various cell subsets active in DTH are commercially available. Strong experimental allergens easily sensitize mice. However, contact sensitivity could not be easily induced by environmental chemicals, and the response was short-lived.

Recently, two groups have reported mouse models capable of identifying so-called weak sensitizers. Both use open challenge to the ear and the increase in ear thickness as the end point for sensitization.

Maisey and Miller (16) prescribed a 2-week induction period with a total of six topical applications to the shaved abdomen and thorax followed a week later by challenge to the ears. They amplified the presentation of the environmental allergens by giving vitamin A acetate in the diet.

Gad et al. (17) developed the mouse ear swelling test (MEST), which prescribes a 4-day induction stage and challenge at day 10. The animals are shaved and tape-stripped on the abdomen at the start of the test. Freund's complete adjuvant is injected into the induction area prior to the first application, and tape stripping and topical application is repeated for 3 additional consecutive days. Gad et al. investigated 10 different strains of mice and found CF-1 and BALB/c mice to be superior to the other strains. The assay was validated by testing 72 materials representing a broad spectrum of chemicals, and 48 of 49 compounds were correctly identified as potential sensitizers, while 23 of 23 compounds were nonsensitizers. Further validation using the technique in different laboratories is necessary.

KEY FACTORS IN INTERPRETATION

Guinea pig allergy test results cannot stand alone. A test giving a questionable result should be repeated, often in a modified form. Much of our interpretation of animal assays must be viewed in light of our knowledge of what happens when other allergens enter human commerce. For example, what can one say about the validity of a guinea pig maximization test that is negative when, in fact, 15% of exposed patients are sensitized (118)? The animal test results should be validated by comparison to other sources of information, such as: (1) limited-use tests on a final formulation prior to marketing; (2) industry, worker, and consumer complaint data; (3) retrospective epidemiological data; (4) monitoring programs such as the Department of Health, Education, and Welfare (HEW)—sponsored investigation of consumer's perceptions of adverse reactions to cosmetic products; and (5) diagnostic test results in dermatological clinics (119). Not until more precise epidemiology data become available will we further refine the interpretation of our animal and human predictive sensitization assays.

The aim of the test used defines its limitations. The guinea pig maximization test establishes to what extent a particular substance has the potential for acting as a contact sensitizer (18). A positive test does not necessarily exclude the substance for use because the stimulus in the test procedure is exaggerated compared to conditions of use. On the other hand, two negative test results from experienced laboratories using the guinea pig maximization test indicates that the substance is unlikely to sensitize. It must be emphasized that this result does not mean that the

"Predictive" Animal Sensitization Assays

substance will never sensitize anyone. For example, lanolin allergy is a clinical problem for leg ulcer patients, even though the batch used by Magnusson and Kligman (18) did not sensitize guinea pigs under experimental conditions. However, recently alkane-diols were identified as the major allergenic constituents of hydrogenated lanolin. These fractions sensitized guinea pigs (120). Hypersensitivity produced in guinea pigs using one of the less-sensitive nonadjuvant techniques (e.g., the open epicutaneous test) is more alarming. This makes it a reasonable approach to use an adjuvant test as a screening procedure and to supplement with a nonadjuvant technique in cases of sensitization.

Positive tests in any of the assays should undergo the rigorous follow-up now employed in human testing, so as to minimize the likelihood of nonspecific positives from the excited skin syndrome (ESS). This entails, at a minimum verification with appropriate concentrations and vehicles, and often obtaining additional controls. Repeating the assay with other techniques and utilization of open challenges (to avoid nonspecific effects of tape) are of great scientific and practical value.

A risk-benefit analysis is appropriate when interpreting results from guinea pig allergy tests. For cosmetic chemicals intended to be used on normal and diseased skin, no substantial risk will be accepted. Many drugs are fully acceptable in spite of sensitizing potential because of their inherent benefit.

The investigator should know the basic chemical purity of the test substance, and the relationship to known allergens (e.g., cross-reactions).

Statistical considerations are involved when extrapolating from a small test population to large numbers of users. Henderson and Riley (121) discussed this issue by drawing attention to the confidence limits. The figures are correct when the test exposure is similar to the use. Many guinea pig allergy tests amplify the exposure to such an extent that negative results indicate a considerable safety margin.

WHAT TO TEST AND NOT TO TEST

A list of priorities can be made. There may not be adequate resources to test all chemicals. Some soaps, dish-washing liquids, and detergents are examples of products that have been used for a long time and by many people without causing allergic contact dermatitis. Unfortunately, some common chemicals, long thought to be "safe," have turned out to be occult allergens, so a high index of suspicion is indicated (108).

New substances intended for skin contact, rubber chemicals, preservatives, drugs, and reactive additives in plastics industry are examples of compounds that should be tested for sensitizing capacity (122).

Consumer groups and regulatory agencies will continue to demand more toxicological evaluation to assure safety of environmental substances. The demand for toxicological data ought not to be a bottleneck that delays development of new products to the benefit of the population. Guinea pig allergy tests are an important tool for safe development of new products when the tests are executed and interpreted by trained people, who at the same time should be able to argue against the possible misuse of the demand for testing and misuse of controversial test results set forth by different interest groups.

REFERENCES

1. Johnson, M.-L. T., Burdick, A. E., Johnson, K. G., Klarman, H. E., Krasner, M., McDowell, A. J. M., and Roberts, J. Prevalence, morbidity and cost of dermatological diseases. *J. Invest. Dermatol.* 73:395–401 (1979).

2. Christoffersen, J. Skin diseases in Denmark. Thesis, Danish Institute of Clinical Epidemiology, Copenhagen (1984).

3. Andersen, K. E., and Maibach, H. I. Drugs used topically. In: *Allergic Reactions to Drugs. Handbook of Experimental Pharmacology*, Vol. 63 (A. L. de Weck and H. Bundgaard, eds.). Springer, Berlin, pp. 313–377 (1983).

4. Rudzki, E., and Zakrzewski, Z. Incidence of contact sensitivity to topically applied drugs as compared with the frequency of their prescription. *Contact Dermatitis* 1:249–250 (1975).

5. Eiermann, H. J., Larsen, W., Maibach, H. I., and Taylor, J. S. Prospective study of cosmetic reactions 1977–1980. *J. Am. Acad. Dermatol.* 6:909–917 (1982).

6. Adams, R. M., and Maibach, H. I. A five year study of cosmetic reactions. *J. Am. Acad. Dermatol.* 13:1062–1069 (1985).

7. Bandmann, H. J., Calnan, C. D., Cronin, E., Fregert, S., Hjorth, N., Magnusson, B., Maibach, H. I., Malten, K. E.,

Meneghini, C. L., Pirila, V., and Wilkinson, D. S. Dermatitis from applied medicaments. *Arch. Dermatol.* 106: 335–337 (1972).

8. Mitchell, J. C., and Maibach, H. I. The angry back syndrome—The excited skin syndrome. *Semin. Dermatol.* 1: 9–13 (1982).

9. Bruynzeel, D. P., and Maibach, H. I. Excited skin syndrome (angry back). *Arch. Dermatol.* 122:323–328 (1986).

10. Prystowsky, S. D., Allen, A. M., Smith, R. W., et al. Allergic contact hypersensitivity to nickel, neomycin, ethylenediamine and benzocaine. *Arch. Dermatol.* 115: 959–962 (1979).

11. Menné, T., Borgan, Ø., and Green, A. Nickel allergy and hand dermatitis in a stratified sample of the Danish female population. An epidemiological study including a statistic appendix. *Acta Derm-Vener., Stockh.* 62:35–41 (1982).

12. Marzulli, F. N., and Maibach, H. I. *Dermatotoxicology.* Hemisphere, Washington, DC (1983).

13. Andersen, K. E., and Maibach, H. I. Contact allergy predictive tests in guinea pigs. *Curr. Probl. Dermatol.*, Vol. 14. Karger, Basel (1985).

14. Maurer, T. *Contact and Photocontact Allergens.* Dekker, New York, p. 74 (1983).

15. Möller, H. Attempts to induce contact allergy to nickel in the mouse. *Contact Dermatitis* 10:65–68 (1984).

16. Maisey, J., and Miller, K. Assessment of the ability of mice fed on vitamin A supplemented diet to respond to a variety of potential contact sensitizers. *Contact Dermatitis* 15:17–23 (1986).

17. Gad, S. C., Dunn, B. J., Dobbs, D. W., Reilly, C., and Walsh, R. D. Development and validation of an alternative dermal sensitization test: The mouse ear swelling test (MEST), *Toxicol. Appl. Pharmacol.* 84:93–114 (1986).

18. Magnusson, B., and Kligman, A. M. *Allergic Contact Dermatitis in the Guinea Pig.* Charles C. Thomas, Springfield, IL (1970).

19. Chase, M. W. Inheritance in guinea pigs of the susceptibility to skin sensitization with simple chemical compounds. *J. Exp. Med.* 73:711–726 (1941).
20. Chase, M. E. The inheritance of susceptibility to drug allergy in guinea pigs. *Trans. NY Acad. Sci., ser. II* 15: 79–82 (1952/1953).
21. Polak, L., Barnes, J. M., and Turk, J. L. The genetic control of contact sensitization to inorganic metal compounds in guinea pigs. *Immunology* 14:707–711 (1968).
22. Parker, D., Sommer, G., and Turk, J. L. Variation in guinea pig responsiveness. *Cell. Immunol.* 18:233–238 (1975).
23. Draize, J. H. *Dermal Toxicity, Appraisal of the Safety of Chemicals in Foods, Drugs and Cosmetics.* Association of Food and Drug Officials in the United States, Austin, TX, p. 46 (1959).
24. Landsteiner, K., and Jacobs, I. Studies on the sensitization of animals with simple chemical compounds. *J. Exp. Med.* 61:643–656 (1935).
25. Voss, J. G. Skin sensitization by mercaptans of low molecular weight. *J. Invest. Dermatol.* 31:273–279 (1958).
26. Prince, H. N., and Prince, T. G. Comparative guinea pig assays for contact hypersensitivity. *Cosmet. Toiletr.* 92: 53–58 (1977).
27. Sharp, D. W. The sensitization potential of some perfume ingredients tested using a modified Draize procedure. *Toxicology* 9:261–271 (1978).
28. Maurer, T., Thomann, P., Weirich, W. G., and Hess, R. The optimization test in the guinea pig. A method for the predictive evaluation of the contact allergenicity of chemicals. *Agents Actions* 5:174–179 (1975).
29. Klecak, G., Geleick, H., and Frey, J. R. Screening of fragrance materials for allergenicity in the guinea pig. I. Comparison of four testing methods. *J. Soc. Cosmet. Chem.* 28: 53–64 (1977).
30. Goodwin, B. F. J., Crewel, R. W. R., and Johnson, A. W. A comparison of three guinea pig sensitization procedures for the detection of 19 reported human contact sentitizers. *Contact Dermatitis* 7:248–258 (1981).

31. Buehler, E. V. Delayed contact hypersensitivity in the guinea pig. *Arch. Dermatol.* 91:171–177 (1965).

32. Buehler, E. V., and Griffith, J. F. Experimental skin sensitization in the guinea pig and man. In: *Animal Models in Dermatology* (H. I. Maibach, ed.). Churchill, Livingstone, New York, Edinburgh, pp. 56–66 (1975).

33. Griffith, J. F., and Buehler, E. V. Prediction of skin irritancy and sensitizing potential by testing with animals and man. In: *Cutaneous Toxicity* (V. A. Drill and P. Lazar, eds.). Academic Press, New York, pp. 155–173 (1977).

34. Marzulli, F. N., and Maguire, H. C., Jr. Validation of guinea pig tests for skin hypersensitivity. In: *Dermatotoxicology* (F. N. Marzulli and H. I. Maibach, eds.). Hemisphere, Washington, DC, pp. 237–250 (1983).

35. Buehler, E. V. Comment on guinea pig test methods. *Food Chem. Toxicol.* 20:494 (1982).

36. Stevens, M. A. Use of the albino guinea pig to detect the skin-sensitizing ability of chemicals. *Br. J. Indust. Med.* 24:189–202 (1967).

37. Klecak, G. Identification of contact allergens: Predictive tests in animals. In: *Dermatotoxicology* (F. N. Marzulli and H. I. Maibach, eds.). Hemisphere, Washington, DC, pp. 193–236 (1983).

38. Fahr, H., Noster, U., and Schulz, K. H. Comparison of guinea pig sensitization methods. *Contact Dermatitis* 2: 335–339 (1976).

39. Andersen, K. E., and Hamann, K. How sensitizing is chlorocresol? Allergy tests in guinea pigs versus the clinical experience. *Contact Dermatits* 11:11–20 (1984).

40. Sato, Y., Katsumura, Y., Ichikawa, H., Kobayashi, T., Kozuka, T., Morikawa, F., and Ohta, S. A modified technique of guinea pig testing to identify hypersensitivity allergens. *Contact Dermatitis* 7:225–237 (1981).

41. Tsuchiya, S., Kondo, M., Okamoto, K., and Takase, Y. Studies on contact hypersensitivity in the guinea pig. The cumulative contact enhancement test. *Contact Dermatitis* 8: 246–255 (1982).

42. Guillot, J. P., Gonnet, J. F., Clement, C., and Faccini, J. M. Comparative study of methods chosen by the Association

fracaise de Normalisation (AFNOR) for evaluation of sensitizing potential in the albino guinea pig. *Food Chem. Toxicol.* 21:795–805 (1983).

43. Brulos, M. F., Guillot, J. P., Martini, M. C., and Cotte, J. The influence of perfumes on the sensitizing potential of cosmetic bases. I. A technique for evaluating sensitizing potential. *J. Soc. Cosmet. Chem.* 28:357–365 (1977).

44. Nater, J. P., and Hoedemaeker, P. J. Histological differences between irritant and allergic patch test reactions in man. *Contact Dermatitis* 2:247–253 (1976).

45. Scheynius, A., Fischer, T., Forsum, U., and Klareskog, L. Phenotypic characterization in situ of inflammatory cells in allergic and irritant contact dermatitis in man. *Clin. Exp. Immunol.* 55:81–90 (1984).

46. Polak, L. Immunological aspects of contact sensitivity. *Monogr. Allergy*, Vol. 15. Karger, Basel, p. 43 (1980).

47. Maguire, H. C., Jr., and Chase, M. W. Studies on the sensitization of animals with simple chemical compounds. Part XIII. *J. Exp. Med.* 135:357–375 (1972).

48. Maguire, H. C., Jr. Estimation of the allergenicity of prospective human contact sensitizers in the guinea pig. In: *Animal Models in Dermatology* (H. I. Maibach, ed.). Churchill Livingstone, London, pp. 67–75 (1975).

49. Andersen, K. E., and Hamann, K. Is Cytox 3522 (10%-methylene-bis-thiocyanate) a human skin sensitizer? *Contact Dermatitis* 9:186–190 (1983).

50. Ziegler, V. Der tierexperimentelle Nachweis allergener Eigenschaften von Industrieprodukten. *Dermatol. Mschr.* 163:387–391 (1977).

51. Polak, L., and Turk, J. L. Studies on the effect of systemic administration of sensitizers in guinea pigs with contact sensitivity to inorganic metal compounds. I. The induction of immunological unresponsiveness in already sensitized animals. *Clin. Exp. Immunol.* 3:245–251 (1968).

52. Doussou, K. G., Sicard, C., Kalopissis, G., Reymond, D., and Schaeffer, H. Method for assessment of experimental allergy in guinea pigs adapted to cosmetic ingredients. *Contact Dermatitis* 13:226–234 (1985).

53. Baer, R. L., and Turk, J. L. Delayed skin reactions. In: *Biochemistry and Physiology of the Skin* (L. A. Goldsmith, ed.). Oxford University Press, New York, pp. 921–937 (1983).

54. Dupuis, G., and Benezra, C. *Allergic Contact Dermatitis to Simple Chemicals*. Dekker, New York (1982).

55. Silberberg, I. Apposition of mononuclear cells to Langerhans cells in contact allergic reactions. An ultrastructural study. *Acta Derm-Vener., Stockh.* 53:1–12 (1973).

56. Stingl, G., Katz, S. I., Shevach, A. M., Rosenthal, A. S., and Green, I. Analogous functions of macrophages and Langerhans cells in the initiation of the immune response. *J. Invest. Dermatol.* 71:59–64 (1978).

57. Landsteiner, K., and Jacobs, J. Studies on the sensitization of animals with simple chemical compounds. Part II. *J. Exp. Med.* 64:625–639 (1936).

58. Godfrey, H. P., and Baer, H. The effect of physical and chemical properties of the sensitizing substance on the induction and elicitation of delayed contact sensitivity. *J. Immunol.* 106:431–441 (1971).

59. Roberts, D. W., and Williams, D. L. The derivation of quantitative correlations between skin sensitization and physicochemical parameters for alkylating agents, and their application to experimental data for sultones. *J. Theor. Biol.* 99:807–825 (1982).

60. Kligman, A. M. The identification of contact allergens by human assay. III. The maximization test -- A procedure for screening and rating contact sensitizers. *J. Invest. Dermatol.* 47:393–409 (1966).

61. Mali, J-W. H., Malten, K., and van Neer, F. C. J. Allergy to chromium. *Arch. Dermatol.* 93:41–44 (1966).

62. Hjorth, N. The allergens. In: *Occupational and Industrial Dermatology* (H. I. Maibach and G. A. Gellin, eds.). Year Book, London, pp. 20–33 (1982).

63. Benezra, C., Sigman, C. C., Perry, L. R., Helmes, C. T., and Maibach, H. I. A systematic search for structure-activity relationships of skin contact sensitizers: Methodology. *J. Invest. Dermatol.* 85:351–356 (1985).

64. Skog, E. Sensitization to p-phenylenediamine. *Arch. Dermatol.* 92:276–280 (1965).
65. Marzulli, F. N., and Maibach, H. I. The use of graded concentrations in studying skin sensitizers: Experimental contact sensitization in man. *Food Cosmet. Toxicol.* 12:219–227 (1974).
66. Hjorth, N., and Trolle Lassen, C. Skin reactions to ointment bases. *Trans. St. John's Hosp. Derm. Soc.* 49:127–140 (1963).
67. Wilkinson, J. D., Hambly, E. M., and Wilkinson, D. S. Comparison of patch test results in two adjacent areas of England. II. Medicaments. *Acta Derm-Vener. Stockh.* 60:245–249 (1980).
68. Chan, P. K., Baldwin, R. C., Parsons, R. D., Moss, J. N., and Hayes, A. W. Kathon biocide: manifestation of delayed contact dermatitis in guinea pigs is dependent on the concentration for induction and challenge. *J. Invest. Dermatol.* 81:409–411 (1983).
69. Hannuksela, M. Rapid increase in contact allergy to Kathon CG in Finland. *Contact Dermatitis* 15:211–214 (1986).
70. Hjorth, N., and Roed-Petersen, J. Patch test reactivity to Kathon CG. *Contact Dermatitis* 14:155–157 (1986).
71. Høvding, G. Cement eczema and chromium allergy. An epidemiological investigation. Thesis, Bergen (1970).
72. Fregert, S., Gruvberger, B., and Sandahl, E. Reduction of chromate in cement by iron sulfate. *Contact Dermatitis* 5:39–42 (1979).
73. Adams, R. M. *Occupational Skin Disease.* Grune & Stratton, New York (1983).
74. Jordan, W. P., Sherman, W. T., and King, S. E. Threshold response in formaldehyde-sensitive subjects. *J. Am. Acad. Dermatol.* 1:44–48 (1979).
75. Lowney, E. D. Dermatologic implications of immunologic unresponsiveness. *J. Invest. Dermatol.* 54:355–364 (1970).
76. Polak, L., and Turk, J. L. Studies on the effect of systemic administration of sensitizers in guinea pigs with contact sensitivity to inorganic metal compounds. I. The induction of immunological unresponsiveness in already sensitized animals. *Clin. Exp. Immunol.* 3:245–251 (1968).

77. Epstein, W. L., Byers, V. S., and Baer, H. Induction of persistent tolerance to urishiol in humans. *J. Allergy Clin. Immunol.* 68:20–25 (1981).

78. Marzulli, F. N., and Maibach, H. I. Further studies of effects of vehicles and elicitation concentration in experimental contact sensitization testing in humans. *Contact Dermatitis* 6:131–133 (1980).

79. Andersen, K. E., Boman, A., Hamann, K., and Wahlberg, J. E. Guinea pig maximization tests with formaldehyde releasers. Results from two laboratories. *Contact Dermatitis* 10:257–266 (1984).

80. Christensen, O. B., Christensen, M. B., and Maibach, H. I. Effect of vehicle on elicitation of DNCB contact allergy in the guinea pig. *Contact Dermatitis* 10:166–169 (1984).

81. Hjorth, N., and Thomsen, K. Differences in sensitizing capacity of neomycin in creams and in ointments. *Br. J. Dermatol.* 80:163–169 (1968).

82. Fischer, T. I., and Maibach, H. I. The thin layer rapid use epicutaneous test (TRUE test), a new patch test method with high accuracy. *Br. J. Dermatol.* 112:63–68 (1985).

83. Andersen, K. E., Hjorth, N., Bundgaard, H., and Johansen, M. Formaldehyde in a hypoallergenic non-woven textile acrylate tape. *Contact Dermatitis* 9:228 (1983).

84. Johansen, M., and Bundgaard, H. Kinetics of formaldehyde release from the cosmetic preservative Germall 115. *Arch. Pharm. Chem. Sci.* 88:117–122 (1981).

85. Chung, C. W., and Giles, A. L. Sensitization potentials of methyl, ethyl, and n-butyl methacrylates and mutual cross-sensitivity in guinea pigs. *J. Invest. Dermatol.* 68:187–190 (1977).

86. Van der Walle, H. B., Klecak, G., Geleick, H., and Benzink, T. Sensitizing potential of 14 mono(meth)acrylates in the guinea pig. *Contact Dermatitis* 8:223–235 (1982).

87. Andersen, K. E., Carlsen, L., Egsgaard, H., and Larsen, E. Contact sensitivity and bioavailability of chlorocresol. *Contact Dermatitis* 13:246–251 (1985).

88. Cronin, E. *Contact Dermatitis.* Churchill Livingstone, Edinburgh (1980).

89. Maibach, H. I., and Prystowsky, S. D. Glutaraldehyde (pentanedial) allergic contact dermatitis. *Arch. Dermatol.* 113:170–171 (1977).

90. Feldmann, R. J., and Maibach, H. I. Regional variation in percutaneous penetration of C-cortisone in man. *J. Invest. Dermatol.* 48:181–183 (1967).

91. Feldmann, R. J., and Maibach, H. I. Penetration of 14C-hydrocortisone through normal skin. *Arch. Dermatol.* 91:661–666 (1965).

92. Macher, E., and Chase, M. W. Studies on the sensitization of animals with simple chemical compounds. XII. The influence of excision of allergenic depots on onset of delayed hypersensitivity and tolerance. *J. Exp. Med.* 129:103–121 (1969).

93. Godfrey, H. P., and Baer, H. The effect of excision of the site of application on the induction of delayed contact sensitivity. *J. Immunol.* 197:1643–1646 (1971).

94. Wester, R. C., and Maibach, H. I. Cutaneous pharmacokinetics: 10 steps to percutaneous absorption. *Drug Metab. Rev.* 14:169–205 (1983).

95. Vadas, M. A., Miller, J. F. A. P., Whitelaw, A. M., and Gamble, J. R. Regulation by the H-2 gene complex of delayed type hypersensitivity. *Immunogenetics* 4:137–153 (1977).

96. Stampf, J. L., Benezra, C., Klecak, G., Geleick, H., Schultz, K. H., and Hausen, B. The sensitizing capacity of helenin, and of two of its main constituents, the sesquiterpene lactones alantolactone and isoalantolactone: A comparison of epicutaneous and intradermal sensitizing methods in different strains of guinea pig. *Contact Dermatitis* 8:16–24 (1982).

97. Andersen, K. E., Boman, A., Vølund, Aa., and Wahlberg, J. E. Induction of formaldehyde contact sensitivity. Dose response relationship in the guinea pig maximization test. *Acta Derm-Vener. Stockh.* 65:472–478 (1985).

98. Menné, T., and Holm, N. V. Nickel allergy in a female twin population. *Int. J. Dermatol.* 2:22–28 (1983).

99. Walker, F. B., Smith, P. D., and Maibach, H. I. Genetic factors in human allergic contact dermatitis. *Int. Arch. Allergy* 32:453 (1967).

100. Menné, T., and Holm, N. V. Hand eczema in nickel sensitive twins. Genetic predisposition and environmental factors. *Contact Dermatitis* 9:289–296 (1983).

101. Calnan, C. D. The climate of contact dermatitis. *Acta Derm-Vener. Stockh.* 44:33–43 (1964).

102. Hjorth, N. *Primula* dermatitis. *Trans. St. John's Hosp. Dermatol. Soc.* 52:207–219 (1966).

103. Turk, J. L., and Parker, D. Sensitization with Cr, Ni, and Zr salts and allergic type granuloma formation in the guinea pig. *J. Invest. Dermatol.* 68:341–345 (1977).

104. Maurer, T., Thomann, P., Weirich, E. G., and Hess, R. Predictive evaluation in animals of the contact allergenic potential of medically important substances. I. Comparison of different methods of inducing and measuring cutaneous sensitization. *Contact Dermatitis* 4:321–333 (1978).

105. Kero, M., and Hannuksela, M. Guinea pig maximization test, open epicutaneous test, and chamber test in induction of delayed contact hypersensitivity. *Contact Dermatitis* 6:341–344 (1980).

106. Maurer, T., Weirich, E. G., and Hess, R. The optimization test in the guinea pig in relation to other predictive sensitization methods. *Toxicology* 15:163–171 (1980).

107. Ritz, H. L., and Buehler, E. V. Planning, conduct and interpretation of guinea pig sensitization patch tests. In: *Current Concepts in Cutaneous Toxicity* (V. A. Drill and P. Lazar, eds.). Academic Press, New York, pp. 25–40 (1980).

108. Roberts, D. W., Goodwin, B. F. J., Williams, D. L., Jones, A. W., and Alderson, J. C. D. Correlation between skin sensitization potential and chemical reactivity for p-nitrobenzyl compounds. *Food Chem. Toxicol.* 21:811–813 (1983).

109. Stadler, J. C., and Karol, M. H. Use of dose-response data to compare the skin sensitizing abilities of dicyclohexylmethane 4,4' diisocyanate and picryl chloride in two animal species. *Toxicol. Appl. Pharmacol.* 78:445–450 (1985).

110. Andersen, K. E. Sensitivity and subsequent "down regulation" of sensitivity induced by chlorocresol in guinea pigs. *Arch. Dermatol. Res.* 277:84–87 (1985).

111. Polak, L., and Scheper, R. J. In vitro DNA synthesis in lymphocytes from guinea pigs epicutaneously sensitized with DNCB. *J. Invest. Dermatol.* 76:133–136 (1981).

112. Hjorth, N. Diagnostic patch testing. In: *Dermatotoxicology*, 2nd ed. (F. N. Marzulli and H. I. Maibach, eds.). Hemisphere, Washington, DC, pp. 267–277 (1983).

113. Weil, C. S., and Scala, R. A. Study of intra- and interlaboratory variability in the results of rabbit eye and skin irritation test. *Toxicol. Appl. Pharmacol.* 19:276–360 (1971).

114. Steinberg, M., Akers, W. A., Weeks, M., McCreesh, A. H., and Maibach, H. I. A comparison of test techniques based on rabbit and human skin responses to irritants with recommendation regarding the evaluation of mildly or moderately irritating compounds. In: *Animal Models in Dermatology* (H. I. Maibach, ed.). Churchill Livingstone, New York, pp. 1–11 (1975).

115. Bruynzeel, D. P., von Blomberg-van der Flier, B. M. E., van Ketel, W. G., and Scheper, R. J. Depression or enhancement of skin reactivity by inflammatory processes in the guinea pig. *Int. Arch. Allergy Appl. Immunol.* 72:67–70 (1983).

116. Andersen, K. E., and Staberg, B. Quantitation of contact allergy in guinea pigs by measuring changes in skin blood flow and skin fold thickness. *Acta Derm-Vener. Stockh.* 65:37–42 (1985).

117. FDA. Good laboratory practice for nonclinical laboratory studies. *Fed. Regul.* 32:427 (1978).

118. Maibach, H. I. Clonidine: Irritant and allergic contact dermatitis assays. *Contact Dermatitis* 12:192–195 (1985).

119. Marzulli, F. N., and Maibach, H. I. Contact allergy: predictive testing in humans. in: *Dermatotoxicology* (F. N. Marzulli and H. I. Maibach, eds.). Hemisphere, Washington, DC, pp. 279–299 (1983).

120. Takano, S., Yamanaka, M., and Okamoto, K. Allergens of lanolin. Part I, II. *J. Soc. Cosmet. Chem.* 34:99–125 (1983).

121. Henderson, C. R., and Riley, E. C. Certain statistical considerations in patch testing. *J. Invest. Dermatol.* 6:227–232 (1945).

122. Magnusson, B., Fregert, S., and Wahlberg, J. E. Determining the allergic contact potential of chemicals. In: *Occupational and Industrial Dermatology* (H. I. Maibach and Gellin, eds.). Year Book, Chicago, pp. 173–185 (1982).

4
Predictive Animal Phototesting

ROBERT L. RIETSCHEL *Ochsner Clinic, New Orleans, Louisiana*

INTRODUCTION

The goal of predictive animal phototesting is to identify potential phototoxic and photoallergenic compounds correctly without falsely labeling a compound as photoactive. The benchmark for any animal methodology is the correct identification of those substances currently noted for phototoxicity (such as psoralens) and photoallergy (such as the salicylanilides).

Phototoxicity

Phototoxicity is generally manifested as an exaggerated sunburn response. The presence of erythema at the site of irradiation without similar erythema at an identically treated but nonirritated site is the end point of a positive phototoxicity test. Several factors are pivotal to successful testing: correct wavelength, proper timing of exposure, vehicle selection, active ingredient penetration, and topical vs. systemic exposure (1).

Regardless of animal model chosen, the proper wavelength of light for phototoxicity testing is a nonerythemagenic dose on normal, stripped, or vehicle-treated skin. Most known compounds are activated in the UVA range (320–400 nm). Doses of 5–15 Joules (J) have proved adequate for elicitation of photoallergy (1), and doses of approximately 1 J are capable of inducing phototoxicity (2). Higher doses of UVA light are often employed in phototoxicity work as UVA light is generally nonerythemagenic in guinea pig animal systems up to 20 J (3). The exposure to light should follow an adequate interval for absorption of the compound, be it topical or systemic. For some compounds, like the psoralens, peak toxicity will occur within 1–2 hr in animals and decrease thereafter whether administered topically or systemically (4). Thus, ultraviolet exposure should occur 1–2 hr after topical or systemic administration. Irradiation after 24–48 hr of open or repeated open patch testing is part of photoallergy evaluation and provides a screen for slowly absorbed or metabolized products that are phototoxic. The onset of phototoxic reactions is characteristically delayed hours to days, and readings should be taken for up to 4 days following exposure (1). The animals most frequently used for phototoxicity testing include the hairless mouse, rabbit, and guinea pig (5,6).

All the subtleties of percutaneous penetration can influence phototoxicity testing. There does not appear to be a universal vehicle that will facilitate this work. Carbowax (polyethylene glycol) virtually eliminated the phototoxicity of methoxsalen, chlorpromazine, and crude coal tar, while hydrophilic emulsions increased detection of phototoxicity (7). These three phototoxic compounds (methoxsalen, chlorpromazine, crude coal tar) provide a good positive control for animal phototoxicity testing systems.

PHOTOALLERGY

Photoallergic compounds became an area of interest when sulfanilamide, phenothiazine, and halogenated salicylanilides were shown to produce photosensitivity (2,8). An animal model was developed by Vinson and Borselli (9) that in various modifications proved useful until recently, when musk ambrette and 6-methylcoumarin produced clinical allergy but were not identified by existing test methodology (3,10).

Detection of photoallergy in animals is similar, in most aspects, to screening for allergic contact dermatitis. A period of time is required for induction of sensitivity, followed by an elicitation procedure. Patch testing for photoallergy generally requires open patches for irridation minutes to several hours after skin contact, whereas routine contact allergy work is done with closed patch tests. During induction several steps are taken to maximize the opportunity for the guinea pig to acquire photoallergy. (The albino guinea pig is currently the most widely used model for this work.) Investigators have increased the penetration of photoactive material through the stratum corneum by sodium lauryl sulfate pretreatment (11), tape stripping (12), and mechanical abrading (3).

The techniques of Jordan (3) and Mauer (13,14) call for a low concentration of substance during the induction phase. High concentrations can alternatively be used during induction to maximize sensitization (5−10%) (12). Adjuvants may be needed to fully develop the potential for photoallergy. This may come from intradermal injections of Freund's complete adjuvant (13,15) or irritation with 20% sodium lauryl sulfate (11). Irradiation with 10−30 J of UVA light is performed 30−60 min after application, and the induction applications are usually applied several times over a period of 2−3 weeks (3,12,15). After a rest period of 1−2 weeks, elicitation is attempted with a range of nonirritant concentrations and 10−20 J of UVA. Some investigators use UVB elicitations as well (13). However, UVB has not generally been needed (3,11) and may be suppressive (16). The two procedures that have most successfully identified the classical photoallergens and the newer "problematic" substances, musk ambrette and 6-methyl coumarin, are those detailed by Ichikawa et al. (15) and Jordan (3). Details of the procedures are given in Table 1. Since Jordan's procedure does not require chemical adjuvants but rather a mechanical removal of stratum corneum, it is unclear why stripping failed to identify 6-methyl coumarin as a photosensitizer in the hands of Ichikawa et al., but the use of the Freund's complete adjuvant did (15). Both techniques appear to be at the forefront of current methodology for animal photoallergy testing.

Experience is being gained with mouse ear swelling methods for photoallergy. At present, standardization lags behind the broader experience with the guinea pig. Cyclophosphamide has been used to increase the induction of sensitivity with the mouse model and several strains have been successfully employed (8), but antecedent UVB distant from sensitization sites significantly depresses induction of photosensitivity, giving this model several idiosyncracies that would favor retention of the guinea pig for now (17). In the guinea pig, UVB at sites distant from chemical applications

Table 1 Technique of Photoallergy Testing

	Jordan (3)	Ichikawa et al. (15)
Animal	Female, albino, Hartley strain guinea pigs, 250–300 gm	Female, albino, Hartley strain guinea pigs, 350–450 gm
UVA light source	F72T12 BL/HO bulbs from National Biological Corp. (Cleveland, OH), with 5-mm polyethylene terephthalate filters, 8 mw/cm^2, induction of 30 J/cm^2	General Electric black light fluorescent tubes filtered through 3-mm window glass, 2.85 mw/cm^2 induction, and elicitation at 10.2 J/cm^2
Timing of UVA	60 min after substance application	30 min after substance application
Skin preparation for induction	Clipped and trimmed, then lightly dermabraded with nylon bristle brush rotating at 13,000 rpm	Shave, Neet epilation, 0.1 ml of Freund's complete adjuvant in 4 nuchal sites at first induction only

Induction concentration	0.1–0.15 ml, 0.25% active ingredient in a vehicle of 1 part dipropylene glycol and 24 parts mineral oil	0.1 ml of 0.1–10% active ingredient in acetone or ethanol 15 min after stripping
Induction frequency	5 days/week for 3 weeks	5 times in 14 days (Freund's used only on first treatment)
Rest period	14 days	14 days
Elicitation preparation	Clipped, 2nd challenge clipped and Neet epilated 1 day after first challenge	Shaved and epilated new area
Elicitation concentration	Applications hourly for 6 hr (repeated the following day for 2nd challenge). Concentration same as induction	0.1 ml of 0.1–10% concentration in acetone or ethanol, 30 min after application 10.2 J UVA
Reading times	24 hr and 2nd 24 hr	24 and 48 hr after light exposure

did not suppress detection, but direct UVB irridation of the patch test site did cause some suppression (16). The exquisite sensitivity of mice to UVB light, leading to systemic immunosuppression, seems to favor the guinea pig in most circumstances.

PERSISTENT LIGHT REACTIVITY

The cause of continuing light sensitivity in the absence of further exposure to a photoallergen has been the cause of wide-ranging speculation as to mechanism (18). A recently developed animal model using Hartley strain male albino guinea pigs, UVA light, and emulsified-Freund's complete adjuvant should help answer some of the lingering questions as to mechanism (19). Antigen persistence suspected by Willis and Kligman (20) now seems unlikely as persistent light reactivity can be induced both with and without antigen (19).

REFERENCES

1. Harber, L. C., and Bickers, D. R. *Photosensitivity Diseases: Principles of Diagnosis and Treatment.* Saunders, Philadelphia (1981).
2. Zaynoun, S. T., Johnson, B. E., and Frain-Bele, W. *Contact Dermatitis* 3:225 (1977).
3. Jordan, W. P., Jr. *Contact Dermatitis* 8:109 (1982).
4. Marzulli, F. N., and Maibach, H. I. In: *Models in Dermatology*, Vol. 2 (H. I. Maibach and N. Lowe, eds.). Karger, New York (1985).
5. Marzulli, F. N., and Maibach, H. I. *J. Soc. Cosmet. Chem.* 21:685 (1970).
6. Kornhauser, A., Wamer, W., and Giles, A. *Science* 217:733 (1982).
7. Kaidbey, K. H., and Kligman, A. M. *Arch. Dermatol.* 110: 868 (1974).
8. De Leo, V. A., and Harber, L. C. In: *Models in Dermatology*, Vol. 2 (H. I. Maibach and N. Lowe, eds.). Karger, New York (1985).
9. Vinson, L. J., and Borselli, V. F. *J. Soc. Cosmet. Chem.* 17:123 (1966).

10. Kaidbey, K. H., and Kligman, A. M. *Contact Dermatitis* 6:161 (1980).
11. Horis, T. *J. Invest. Dermatol.* 67:591 (1976).
12. Kochevar, I. E., Zalar, G. L., Einbinder, J., and Harber, L. C. *J. Invest. Dermatol.* 73:144 (1979).
13. Mauer, T., Weirich, E. G., and Hess, R. *Br. J. Dermatol.* 103:593 (1980).
14. Mauer, T. *Contact and Photocontact Allergens, A Manual of Predictive Test Methods.* Marcel Dekker, New York, 1983.
15. Ichikawa, H., Armstrong, R. B., and Harber, L. C. *J. Invest. Dermatol.* 76:498 (1981).
16. Morrison, W. L., Parrish, J. A., Woehler, M. E., and Bloch, K. J. *Br. J. Dermatol.* 104:161 (1981).
17. Granstein, R. D., Morrison, W. L., and Kripke, M. L. *J. Invest. Dermatol.* 80:158 (1983).
18. Wilkinson, D. S. *Br. J. Dermatol.* 74:302–306 (1962).
19. Katsumura, Y., Tanaka, J., Ichikawa, H., et al. *J. Invest. Dermatol.* 87:330–333 (1986).
20. Willis, I., and Kligman, A. M. *J. Invest. Dermatol.* 51:385–394 (1968).

5
Human Predictive Phototesting

EDWARD A. EMMETT* and MARCIA A. LEVY *Johns Hopkins University, Baltimore, Maryland*

INTRODUCTION

Oscar Raab first reported on a biologically adverse combined effect of light and chemical in 1900 when he observed that the time taken for an acridine dye to kill paramecia varied with the amount of sunlight in the laboratory (1). In 1913, Frederich Meyer-Betz demonstrated that local cutaneous responses with marked edema followed the systemic administration of hematoporphyrin and subsequent light exposure (2).

The classic experiments of drug photosensitivity reactions in humans were conducted by Stephan Epstein in the late 1930s. Epstein gave intradermal injections of sulfanilamide (an agent suspected of causing photosensitivity reactions in treated patents) to six subjects, who were then exposed to broad-band

---

**Present affiliation*: National Occupational Health and Safety Commission, Sydney, Australia.

ultraviolet radiation. All subjects developed erythema and edema followed by pigmentation within several days of exposure. Ten days later, however, two of the six tested subjects developed a new and different reaction—an inflammatory, urticarial response accompanied by an intense pruritus, which persisted for 10—14 additional days. When Epstein subsequently reinjected and irradiated these two subjects, they developed a similar urticaria within hours of the retest. He felt that these reactions were quite distinct from those demonstrated by the other four subjects and coined the term "photoallergy" to distinguish them from the other, "primary photosensitivity" reactions (3).

These two major types of photosensitivity are still distinguished, now generally by the terms photoallergy and phototoxicity. The clinical and immunological characteristics of these two reactions have been reviewed (4). Phototoxicity has been recognized to be a chemically induced increased reactivity of a target tissue to ultraviolet and/or visible radiation on a nonimmunological basis. Phototoxic reactions occur only on those parts of the skin exposed to radiation. There is a direct dose-response relationship between the intensity of the reaction and both the concentration of the chemical in the target tissue and the amount of radiation of the appropriate wavelength to which the target tissue is exposed (5). Phototoxic reactions in humans often result in erythema and edema followed by pigmentation, but other types of reactions are possible; including bullae (6), papular reactions (7), onycholysis (8), and other changes. Ocular damage, including keratoconjunctivitis (9) and possibly cataract formation, may occur (10).

In all phototoxic reactions adequately characterized to date, a critical first event in the generation of the photoreaction has been the absorption of radiation by the chemical or one or more of its metabolites producing an excited state. This is in accord with the Grotthus-Draper Law (or the First Law of Photochemistry) which states that to produce a photochemical or photobiological change, radiation must be absorbed. The photochemical reactions responsible for phototoxicity appear to be of two main types. In type I reactions, also known as free-radical or direct reactions, the excited state of the photosensitizing molecule reacts directly with a biomolecule by electron or hydrogen atom transfer to give a semireduced free-radical form of the sensitizer and a semioxidized free-radical form of the biomolecule. In a type II or singlet oxygen reaction, the excited state of the photosensitizing molecule reacts with a ground-state sensitizer to generate a singlet excited state of oxygen, which typically reacts very rapidly to oxidize various biomolecules. The chemistry of these reactions, however, can be very complex, and the intermediate steps are not well understood (11).

In photoallergic reactions there is a chemically induced increased reactivity of target tissue to UV and/or visible radiation on an immunological basis. Cell-mediated immunity has been responsible in all examples adequately characterized to date. Radiation appears to play a role in the generation of the complete allergen in the skin (12). In some instances such as sulfanilamide, the potential photosensitizer may be converted to a photodecomposition product that is a potent ordinary contact sensitizer (13). In other instances, such as tetrachlorosalicylanide (14,15), chlorpromazine (16), and Jadit (17), photosensitization results in covalent binding between the photoallergic compound and skin protein. Thus, for both phototoxic and photoallergic reactions, the absorption of radiation by the photosensitizing chemical is the inciting step in the biological reaction.

For each phototoxic or photoallergic reaction, there is a characteristic action spectrum, the relationship between the intensity of the response and the wavelength of the inciting radiation. The action spectrum of the response generally corresponds quite well with the absorption spectrum of the responsible chromophore; However, some discrepancies may occur. These may result from (1) binding of the photosensitizer to cellular components with a shift in the absorption spectrum, (2) transformation of the photosensitizer by metabolic processes, or (3) attenuation of the absorption spectrum by the refractive indices, chromophore distributions, and light scattering properties of the skin (12,18).

Substances that cause phototoxic or photoallergic reactions absorb in the UV and/or visible spectrum; thus a review of the absorption spectrum should be the first step in the evaluation of potential photosensitizers. Second, it is critical in predictive human phototesting to ensure that tested individuals are exposed to sufficient radiation of those wavelengths encompassed in the action spectrum. For practical purposes, almost all such reactions take place in the UV-A,* but other spectral regions may be involved (19).

---

*The ultraviolet spectrum is conveniently divided into:
   UV-A: 315–400 nm wavelengths
   UV-B: 280–315 nm wavelengths
   UV-C: 200–280 nm wavelengths
Sunlight at the earth's surface contains virtually no radiation shorter than 290 nm.

Although most attention has been given to identifying phototoxins and photoallergens, there are other possible mechanisms whereby a chemical might induce photosensitization in humans. Disease characterized by photosensitization can be induced: e.g., the overproduction of endogenous porphyrin photosensitizers in hexachlorobenzene-induced porphyria cutanea tarda (20); the induction of pellagra, a vitamin deficiency in which photosensitivity can occur as a result of drugs such as INH; or the production of depigmented and thus poorly protected skin from exposure to p-tertiary butylphenol and a number of other agents (21). Such types of photosensitivity may not necessarily be apparent from the standard human predictive phototesting procedures to be described.

PREDICTIVE PHOTOTESTING BY TOPICAL APPLICATION

The Photomaximization Test

Predictive testing to detect photoallergens, analogous to predictive testing to detect ordinary contact allergens, has generally used a repeated insult patch technique. A prototypical test is the photomaximization test (22,23), which entails exaggerated exposures to both chemical and ultraviolet radiation.

For the induction phase, the test chemicals are applied evenly to 2.5-cm squares of skin. The sites are then covered by nonwoven cotton cloth and sealed to the skin with occlusive tape. Twenty-four hours later the patches are removed, the skin wiped dry, and the site irradiated. After a rest period of 48 hr, a similar application with subsequent irradiation is performed. The sequence is repeated for a total of six exposures over 3 weeks. Irradiation is performed using three minimal erythema doses (MEDS) from a 150-watt xenon arc solar simulator. This irradiation source has a continuous spectrum from 290 to 410 nm, which resembles the spectrum of midday summer sunlight at 40°N latitude, particularly in the UVB and shorter UVA (24). The MED, the radiation dose that produces minimal but uniform erythema with a clear margin 24 hr after irradiation, is determined beforehand individually on each participating subject, from the reaction to a series of doses given at 25% increments.

Subjects are challenged after a rest period of 14 days by a single 24-hr application of the test compound to a previously untested skin site, followed by exposure to 4 J/cm^2 UVA. The exposure source is the same solar simulator radiation used for induction filtered through a 2-mm Schott WG345 filter. Similar applications of test compounds are made to control sites, which are sealed and kept covered with three to four layers of opaque adhesive tape. The sites are then examined 48 and 72 hr after irradiation. Development of erythema and edema or a vesicular dermatitis in the irradiated but not the unirradiated sites signifies the induction of photocontact sensitivity. Each substance is usually examined in a separate panel of 25 subjects. For the induction phase, the agent is diluted at 5% concentrations of test agent in petrolatum, and 10 µl of the petrolatum solution is applied per cm^2 of skin. For the challenge phase, the test agent is applied at a 1% concentration in petrolatum.

TOPICAL PHOTOTOXICITY ASSAYS

Assays for topical phototoxicity are generally more straightforward than those for potential photoallergenicity. The substance is applied to the skin in an appropriate vehicle, irradiated after a period with an appropriate light source, and examined later. A control, consisting of a nonirradiated application, serves to exclude non-light-dependent reactions. Reported test details, however, vary rather widely.

Burdick (25) and Marzulli and Maibach (26) apply 0.05 ml of test solutions to a 5 × 12 cm area of forearm skin, tape-stripped to glistening, allow the agent to remain undisturbed for 5 min and subsequently irradiate with a predominantly UVA source. Readings are made at 24 and 48 hr.

Kaidbey and Kligman (27) deliver 50 µl of test material to a 3.2-cm^2 area of untanned midback skin (5 µl/cm^2) and spread it uniformly with a glass rod. The sites are covered with nonwoven cloth and overlapping strips of occlusive tape. After 6 hr the patches are removed, excess material wiped from the skin, and the site irradiated using 17.5 J/cm^2 of UVA from a 150-watt Xenon arc solar simulator with a Schott WG-345 filter to remove UVB. The test is read immediately and after 24 and 48 hr. If no reaction occurs, the UVA dosage is increased to about 28.5 J/cm^2. Sometimes agents are also applied to skin that has been scarified by using a 30-gauge needle to scratch 10 parallel slits in each of two directions.

With these methods, positive reactions may be characterized by burning, smarting, erythema, edema, wheal and flare, blistering, and pigmentation. With different phototoxic agents different types of reactions may be seen. For example, with coal tar (29) or amyl-dimethyl orthoaminobenzoic acid (30), a severe burning sensation is observed within a few minutes of commencing exposure, followed by erythema and flare, which fade within a few hours. Several hours later a sunburn type of response is seen. In the case of the furocoumarins, the appearance of erythema may be delayed for 48–72 hr, but this may be followed by severe blistering (29). With some agents, particularly xanthine and thiazine dyes (29), phototoxicity is observed only when testing is performed on scarified skin. Stripping of the epidermis was necessary to show that griseofulvin and tolbutamide were phototoxic after topical application, but was not necessary for chlorpromazine and sulfanilamide (31). Such techniques presumably increase the percutaneous absorption of the potential photosensitizer but could also increase the resultant inflammatory response (32).

VARIABLES IN PREDICTIVE HUMAN TOPICAL PHOTOTESTING

We know that a considerable number of factors determine our success in inducing allergic contact sensitization and irritant reactions in humans. These include the concentration of the agent, choice of vehicle, frequency of application, amount of application, type of covering, and state of the skin (scarification, treatment with sodium lauryl sulfate, etc.). Additional variables associated with the irradiation include the wavelength composition, exposure duration and intensity, number of exposures, time between application and exposure, and time of reading the reactions, which could also be expected to play an important role in determining the experimental outcome. We already have good indications that certain chemicals behave quite differently in the commonly used tests, presumably reflecting in part pharmacokinetic considerations. Thus, it seems unlikely that a uniform "cookbook" approach to predictive phototesting will be consistently rewarding. However, few systematic studies of many potentially important variables have been undertaken, so that the understanding which would be necessary to make sensible, mechanistically based decisions on techniques is sadly lacking.

In this section we will indicate some of the considerations that should enter into the design of a predictive study, acknowledging that this science is embryonic.

Source and Identity of Test Substance

In general, the test substance is provided by the sponsor, together with whatever information is available on irritancy and toxicity. Some caution is necessary with this information. There is a recent report in the literature of an incorrect description in the manufacturer's declaration of a substance submitted for photopatch testing (33). One of us has had personal experience with a similar situation where a substance causing photosensitivity in industrial workers was wrongly identified on the material safety data sheet provided by the manufacturer. Additionally, potential photosensitizers may contain impurities, and degradation and/or photodegradation products may be formed during storage. Suspect photosensitizers should all be considered to be unstable in light and should be appropriately stored in brown glass bottles covered with aluminum foil.

Method of Application

Predictive phototest techniques have generally involved application of a test compound to the skin with a covering of nonwoven or woven cotton and occlusive tape. A diagnostic patch test method currently in vogue uses small concave chambers that are tightly adherent to the skin. The tight occlusion of the chamber leaves a clearly discernible ring on the skin surface, providing assurance of proper contact (34), but the use of such technique in predictive phototesting has only recently been reported (35).

It is often most convenient to perform predictive phototesting on the back or lumbar area. There are profound regional differences in the permeability of skin (36) and also in the reaction to ultraviolet radiation (37). Marzulli and Maibach (38) found a higher proportion of phototoxic responses to oil of Bergamot on unstripped scrotal and neck skin than on forearm skin. Testing on the scrotal skin, which is highly permeable, has been suggested (38), but seems unlikely to become popular. Predictive phototesting is usually performed on fair-skinned individuals because of the lower necessary exposure doses and the ease of observing erythema.

Concentration and Vehicle

The choice of concentration and vehicle for testing is important. The amount of test compound delivered to the target cell (presumably the living layers of the epidermis and dermis) should be sufficient to produce a reaction, yet avoid irritation in the unirradiated control sites. Systemic toxicity must also be avoided, although this is rarely, if ever, an issue because of the small amount of test substance employed. Kaidbey and Kligman (39) found the results of phototoxicity assays with coal tar, methoxalen, and chlorpromazine strongly influenced by vehicle choice. No single base produced optimal effects for all chemicals. Polyethylene glycol was shown to be a poor vehicle, while emulsion-type creams were generally more active than petrolatum. It would seem prudent with novel test compounds to use several vehicles, or at least an experimental vehicle (such as alcohol) that is likely to release the test compound.

There is a dearth of information as to the optimal amount of test agent to apply.

Interval Between Application of Test Material and Irradiation

This seemingly important variable has been too little studied. With many substances, considerable time is necessary for absorption. In the case of Bergamot, humans will react to irradiation 1–2 hr after application but not after 24 hr (38), whereas a 26-hr delay has proven quite satisfactory for the induction and elicitation of photoallergy and phototoxicity of many agents (19, 22).

Radiation Source and Exposure

Factors of importance include the spectral distribution, filtration, intensity, duration, uniformity, and method of measurement of single exposures, and the method of fractionation and total number of repeated exposures. With agents for which the action spectrum is unknown, a broad-band UV radiation source should be used. The output should be carefully monitored; photodosimeters are now available that can be used together with a detector in the irradiation field to ensure correct dose delivery and uniformity of the irradiation field. The reader is referred to several excellent sources on instrumentation and measurement (12, 40–42).

Number of Subjects

In the case of human topical phototoxicity assays, it is generally considered that only a small number of subjects are necessary because of the relative uniformity of responses given similar chemical dosage and irradiation to the target organ. This assumption is not necessarily correct; for example, in the case of substances whose metabolites are phototoxins, variation in genetic susceptibilities within populations might lead to situations where a larger test panel was desirable. As an example where metabolism is important, a number of metabolites of chlorpromazine are more phototoxic than the parent compound (43).

With photoallergy we are considering an event, dependent on a sensitization process, which may occur in only a few exposed individuals. We would therefore like to include enough persons on any test panel to have some likelihood of detecting reasonably potent photoallergens. Computation of this number is particularly difficult because the agents tested may be used in a large general population to which we would like to meaningfully extrapolate results. Mathematical considerations in such extrapolation have been addressed by Henderson and Riley (44). If there are no reactions in a test population of 200 random subjects, the upper 95% confidence limit for the general population would be reactions in 15 of every 1000 individuals. Further, if 1 of 200 subjects in a test population becomes sensitized, a test population of 10,000 subjects might show from 1 to 275 sensitized at a 95% confidence level. To date, panels for predictive human photoallergy testing generally have been much smaller than 200. To increase the predictive ability of these tests, investigators have relied on maximizing the attendant circumstances that encourage the development of photoallergy, such as the use of relatively large exposures to chemical and light, frequently repeated applications, and so forth (45), rather than on increasing the size of the test panel. There has been little systematic study of the validity of this approach.

VARIATIONS OF TESTING TECHNIQUES

Clinical Trials

Blank et al. (46) and Frost et al. (47) have described the use of a clinical trial to detect photosensitivity from orally administered drugs. Subjects took either a drug at usual daily doses or identical placebo capsules containing lactose, for 1 week. At the end of this time, they were taken on a boat cruise for 5–6 hr where they were exposed to sunlight while wearing bathing suits.

At 1, 3, 6, 24, and 48 hr and at 1 week after exposure, consensus judgments were made by three dematologists as to whether the skin reactions were normal or abnormal. Abnormal reactions were graded from + to ++++ on the basis of unusually accentuated erythema or violaceous color; the presence of parasthesias including tingling and burning sensations, or the appearance of edema, urticaria, vesicles, or papules. Although all participants were said to develop some erythema as a normal response to sun exposures, the authors stated that abnormal reactions could be detected with ease. In these studies demeclocycline was quite phototoxic, whereas methacycline and the placebo were associated with little clinical phototoxicity and doxycycline produced only occasional phototoxicity.

The relative advantages of such clinical trials compared with testing of the skin of volunteers to controlled light exposures under laboratory conditions have not been established. There is little in the literature to indicate whether such clinical trials are effective in identifying mildly phototoxic drugs.

Testing with Preirradiated Substances

As photodecomposition products may be responsible for reactions, testing with preirradiated substances may be performed (48). The usefulness of this procedure has not yet been established.

Intradermally Administered Test Substances

Epstein's early studies utilized intradermal testing of sulfanilamide (3). Intradermal administration has been effective in demonstrating phototoxicity of a variety of drugs including tetracyclines, chlorthiazides, sulfonamides, and others (31). With such testing, irradiation should follow immediately or shortly after the administration of the test drug. For example, within 6 hr after injection, responses to irradiation of demeclocycline injected sites are greatly decreased (31).

THE PLACE OF HUMAN PREDICTIVE TESTING FOR PHOTOSENSITIVITY

Despite the availability of experimental methods to identify many if not most photosensitizers, outbreaks of photosensitivity to novel agents continue to occur. In addition, the photosensitivity generally is recognized only after marketing, thus highlighting the need for predictive tests (22).

Since phototoxicity is often dependent on rather basic photobiological reactions, there is a wide variety of model systems that can be used to predict whether this reaction may occur in humans. These include biochemical, organelle, unicellular, organ culture, simple organism, and whole mammalian systems. These approaches have been summarized by Harber (32) and Emmett (5). Whole mammals offer particular advantages over simpler systems since they appropriately take into account percutaneous and gastrointestinal absorption and distribution within the layers of the skin, as well as cutaneous and extracutaneous metabolism and inactivation.

Phototoxicity testing is often performed first in experimental animals and later in humans. Clearly, human testing can most accurately duplicate the actual circumstances in which photosensitivity may occur in humans.

As the induction of photoallergy requires a functioning immune system as well as the presence of skin, at our present state of knowledge predictive testing for photoallergy requires whole animals. Experimental photoallergy has been induced in rodents, notably guinea pigs, and in humans. Unfortunately, no single experimental animal technique has yet been described to identify all photoallergens. Human testing can result in the induction of photoallergy to a number of agents. For example, using a photomaximization test, Kaidbey and Kligman (49) induced photoallergy to 3,3',4',5-tetrachlorosalicylanilide, commercial tribromosalicylanilide, 3,5-dibromosalicylanilide, 4,5-dibromosalicylanide, Jadit, bithional, 6-methylcoumarin, and chlorpromazine, but not to musk ambrette or sulfanilamide. However, as persistent light reactions have apparently been induced in experimental subjects being tested for photoallergenicity (50,51), this testing should be done judiciously. In addition, as no universally accepted protocols for predictive phototesting exist at this time, the circumstances of use, chemical structure, and likely mode of action of the potential photosensitizer must remain the primary considerations in such an undertaking.

REFERENCES

1. Raab, O. *Z. Biol.* 39:524 (1900).
2. Blum, H. F. *Photodynamic Action and Diseases Caused by Light.* Hafner, New York, p. 211 (1964).
3. Epstein, S. *J. Invest. Dermatol.* 2:43 (1939).
4. Harber, L. C., and Baer, R. L. *J. Invest. Dermatol.* 58: 327 (1972).
5. Emmett, E. A. *Photochem. Photobiol.* 30:429 (1979).
6. Birmingham, D. J., Key, M. M., Tublick, G. E., and Perone, V. B. *Arch. Dermatol.* 83:73 (1961).
7. Frost, P., Weinstein, G. D., and Gomez, E. C. *J.A.M.A.* 216:326 (1971).
8. Frank, S. B., Cohen, H. J., and Miukin, W. *Arch. Dermatol.* 103:520 (1971).

9. Emmett, E. A. *Contact Dermatitis* 3:245 (1977).
10. Parrish, J. A., Stern, R. S., Pathak, M. A., and Fitzpatrick, T. B. *The Science of Photomedicine* (J. D. Regan and J. A. Parrish, eds.). Plenum Press, New York, p. 595 (1982).
11. Spikes, J. D. *Photoimmunology* (J. A. Parrish, M. L. Kripke, and W. L. Morison, eds.). Plenum Press, New York, p. 23 (1983).
12. Harber, L. C., and Bickers, D. R. *Photosensitivity Diseases*. Saunders, Philadelphia, p. 33 (1981).
13. Burckhardt, W., and Schwarz-Speck, M. *Schweiz. Med. Wochenschr.* 87:954 (1957).
14. Kochevar, I. E., and Harber, L. C. *J. Invest. Dermatol.* 68:151 (1977a).
15. Kochevar, I. E. *Photochem. Photobiol.* 30:437 (1979).
16. Jung, E. G. *Arch. Klin. Exp. Dermatol.* 237:501 (1970).
17. Jung, E. G., Hornke, J., and Hajdu, P. *Arch. Klin. Exp. Dermatol.* 233:287 (1968).
18. Anderson, R. R. *Photoimmunology* (J. A. Parrish, M. L. Kripke, and W. L. Morison, eds.). Plenum Press, New York, p. 61 (1983).
19. Emmett, E. A. *Arch. Dermatol.* 110:249 (1974).
20. Schmid, R. *N. Engl. J. Med.* 263:397 (1960).
21. Gellin, G. A., Possick, P. A., and Perone, V. B. *J. Invest. Dermatol.* 55:190 (1970).
22. Kaidbey, K. H., and Kligman, A. M. *Contact Dermatitis* 6:161 (1980).
23. Kaidbey, K. H. *Dermatoxicology 2* (F. Marzuli and H. Maibach, eds.). Hemisphere, New York, p. 405 (1983).
24. Berger, D. S. *J. Invest. Dermatol.* 53:192 (1969).
25. Burdick, K. H. *Arch. Dermatol.* 93:424 (1966).
26. Marzulli, F. N., and Maibach, H. I. *J. Soc. Cosmet. Chem.* 21:695 (1970).
27. Kaidbey, K. H., and Kligman, A. M. *J. Invest. Dermatol.* 70:149 (1978).

28. Frosch, P. J., and Kligman, A. M. *Contact Dermatitis* 2: 314 (1976).
29. Kaidbey, K. H. *Safety and Efficacy of Topical Drugs and Cosmetics* (A. M. Kligman and J. J. Leyden, eds.). Grune & Stratton, New York, p. 213 (1982).
30. Emmett, E. A. *Arch. Dermatol.* 113:770 (1977).
31. Kligman, A. M., and Breit, R. *J. Invest. Dermatol.* 51:90 (1968).
32. Harber, L. C. *J. Invest. Dermatol.* 77:65 (1981).
33. Bruze, M. *Photodermatology* 1:199 (1984).
34. Pirila, V. *Contact Dermatitis* 1:48 (1975).
35. Kavli, G., Raa, J., Johnson, B. E., Volden, G., and Haugsbo, S. *Contact Dermatitis* 9:257 (1983).
36. Feldman, R., and Maibach, H. *J. Invest. Dermatol* 54:399 (1970).
37. Olson, R. L., Sayre, R. M., and Everett, A. M. *Arch. Dermatol.* 93:211 (1966).
38. Marzulli, F. M., and Maibach, H. I. *Models in Dermatology 2* (H. I. Maibach and N. J. Lowes, eds.). Karger, New York, p. 349 (1985).
39. Kaidbey, K. H., and Kligman, A. M. *Arch. Dermatol.* 110:868 (1974).
40. Task Force on Photobiology, Ultra-violet light sources. *Arch. Dermatol.* 109:833 (1974).
41. Landry, R. J., and Anderson, F. A. *J. Natl. Cancer Inst.* 69:155 (1982).
42. Kochevar, I. E., and Anderson, R. R. *Photoimmunology* (J. A. Parrish, M. L. Kripke, and W. L. Morison, eds.). Plenum Press, New York, p. 51 (1983).
43. Ljunggren, B., and Moller, H. *J. Invest. Dermatol.* 68:313 (1977).
44. Henderson, C. R., and Riley, E. C. *J. Invest. Dermatol.* 6:227 (1945).
45. Kligman, A. M. *J. Invest. Dermatol.* 47:369 (1966).

46. Blank, H., Cullen, S. I., and Catalano, P. *Arch. Dermatol.* 97:1 (1968).
47. Frost, P., Weinstein, G., and Gomez, E. *JAMA* 216:326 (1971).
48. Herman, P. S., and Sams, W. M. *Soap Photodermatitis.* Charles C. Thomas, Springfield, IL, p. 47 (1972).
49. Kaidbey, K. H., and Kligman, A. M. *Contact Dermatitis* 6:161 (1980).
50. Willis, I., and Kligman, A. M. *J. Invest. Dermatol.* 51:385 (1968).
51. Willis, I., and Kligman, A. M. *J. Invest. Dermatol.* 51:378 (1968).

6
Measurement of Cutaneous Blood Flow

DIRK B. ROBERTSON *Emory University School of Medicine, Atlanta, Georgia*

HOWARD I. MAIBACH *University of California, San Francisco, California*

INTRODUCTION

The measurement of blood flow in the cutaneous microcirculation of humans is of considerable interest to a diverse group of investigators. Multiple techniques have been devised, including the use of radioactive tracers, thermal uptake from heated probes, photopulse plethysmography (PPG), and laser Doppler velocimetry (LDV). The ideal method would provide continuous measurement of cutaneous blood flow; it would not disturb the state of the circulation and would be indefinitely repeatable; it would also provide absolute values independent of local tissue characteristics and would respond instantaneously to changes (1). The noninvasive optical techniques of PPG and LDV have been extensively utilized within the last decade and provide essentially instantaneous continuous monitoring of cutaneous blood flow. The major disadvantage is that these methods are sensitive to changes in the microvascular bed geometry and optical properties of the skin and therefore require calibration against another method of flow measurement, such as the xenon washout technique, to provide absolute values of cutaneous blood flow. Fortunately, many investigations are not dependent on determining absolute flow measurements, but rather monitor relative perfusion levels to observe how basal flow values are altered by various types and degrees of stress (2). In these instances noninvasive optical techniques come close to providing an ideal method of monitoring cutaneous microcirculation.

Although the technology utilized in the development of these methods is sophisticated, the conceptual basis behind the determinations is straightforward. It is at this basic level that the fundamental concepts and ideas behind the instrumentation of LDV and PPG are presented to enable the reader to obtain a general overview of these methods (3). More extensive details and complete theoretical background regarding these techniques may be found in the references (1, 4–12).

EXPERIMENTAL TECHNIQUES

Laser-Doppler Velocimetry

The LDV technique is based on measurement of the Doppler frequency shift in monochromatic laser light which is backscattered from moving red blood cells. Because the red cells are moving at different velocities, the single laser output frequency is shifted to a spectrum of different frequencies and then measured in terms of broadening of the spectrum. Initial observations on the feasibility of utilizing this system for the measurement of blood flow in the cutaneous microcirculation were made by Stern (4). Subsequent development and design changes resulted in a practical clinical instrument based on the above principles (5).

With this method (Fig. 1), light is transmitted from a continuous-wave laser source in the instrument to the skin via a fiber-optic fiber, where it is directed into the dermis, penetrating to a depth of 1–1.5 mm. The LD 5000 Capillary perfusion monitor of Medpacific (Seattle, WA) and the Periflux PF3 flowmeter of Medex (Hillard, OH) use a helium-neon laser that provides monochromatic light with a wavelength of 632.8 nm, while the Laserflo BPM 403 blood perfusion monitor of TSI (St. Paul, MN) uses a solid-state laser diode that provides coherent light with a wavelength of 780 ± 20 nm. As the incident radiation enters the skin,

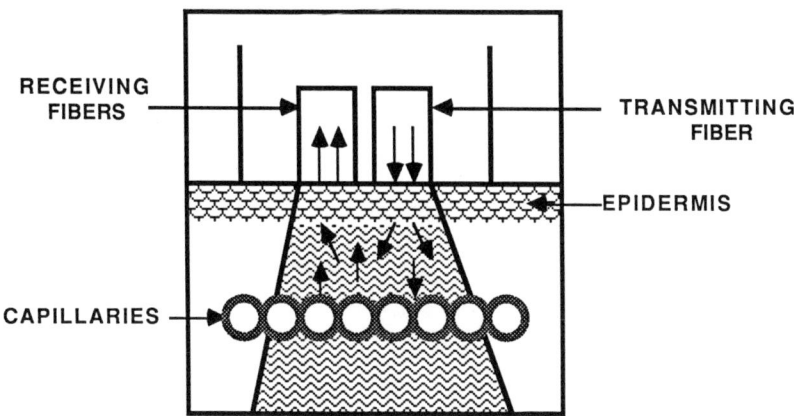

Figure 1 Laser Doppler velocimetry (LDV). Light from a helium-neon laser is transmitted to the skin via an optical fiber. The backscattered light consists of radiation at the same frequency as the incident source and radiation that has been Doppler shifted as a result of reflection from moving red blood cells. The LDV detects the frequency shifted signal and derives an output proportional to the number of erythrocytes times their velocity in the cutaneous microcirculation.

it is multiply scattered and reflected by both stationary tissue and mobile red blood cells. The former backscatter radiation at the incident frequency, the latter reflect light that has been frequency-shifted by an amount proportional to the product of their number multiplied by their velocity. Available devices utilize either a second fiberoptic fiber, in the LD 5000 capillary perfusion monitor of Medpacific, a second and third fiberoptic fiber in the Periflux PF3 flowmeter of Medex, or multiple fiberoptic fibers in the Laserflo BPM 403 blood perfusion monitor of TSI, to return the light to the instrument. Physical support for the fibers at the point where they interface with the skin is provided by a small cylindrical probe, which is attached to the surface with double-sided adhesive tape. The returning radiation is directed on a photodetector and the frequency-shifted component is separated, amplified and displayed as a fluctuating voltage (Fig. 2). As the frequency-modulated portion is directly related to the product of the number of incident light-scattering erythrocytes

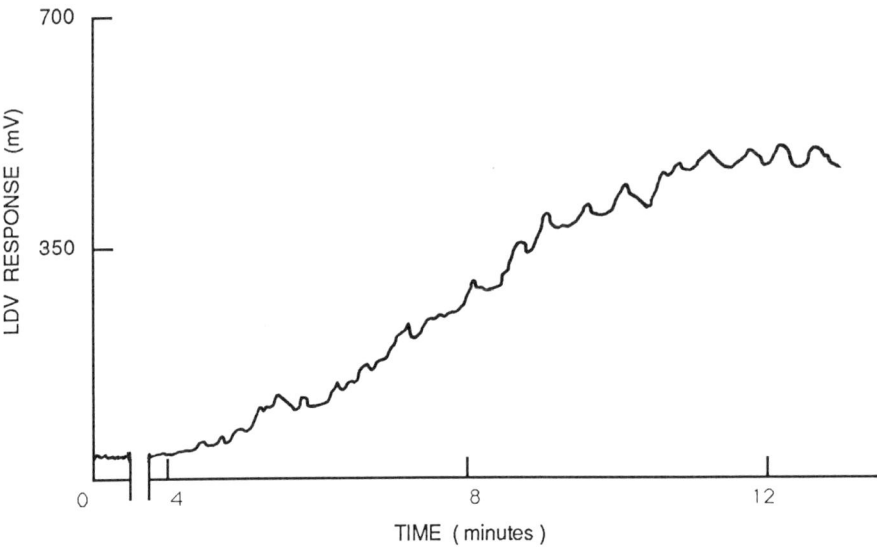

Figure 2 Early time portion of a LDV response curve. The site of application was the ventral forearm, with hexyl nicotinate (200 mM) in a propylene glycol/isopropanol vehicle as the vasodilator.

and their velocity, the flow signal is a sensitive indicator of local cutaneous perfusion and provides a real-time monitor of changes taking place in the superficial microcirculation. The Periflux PF3 flowmeter, which uses two light-receiving fibers, does so to take advantage of a differential analysis procedure, thus permitting improvement in the signal-to-noise ratio (8,9). The TSI Laser-flo blood perfusion monitor utilizes digital signal processing by a microprocessing unit that eliminates drift, offset, and calibration corrections associated with analog circuits and allows for relative measurements of blood flow, velocity, or volume.

Photopulse Plethysmography

The PPG technique utilizes an infrared light-emitting diode (LED), which generates radiation in the frequency range 800–940 nm. At these wavelengths cutaneous tissue is essentially transparent, whereas hemoglobin molecules absorb strongly in this spectral

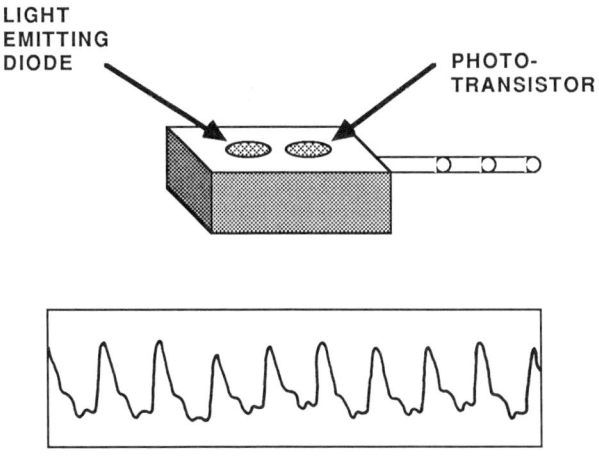

NORMAL ARTERIAL SIGNAL

Figure 3 Photopulse plethysmography (PPG). This method uses an infrared light emitting diode to determine changes in microvascular blood volume passing through the dermal capillaries by measuring the amount of incident radiation absorbed. The signal appears as a series of pulsations.

range. Therefore, when light from the LED is directed down into the skin, the proportion backscattered is attenuated according to the amount of hemoglobin present. The process by which the blood itself in the microcirculation causes attenuation of the radiation is complex, but essentially the more blood there is, the greater the attenuation (12). The penetration depth of the light from the PPG probe is similar to that of the LDV, with a depth of 1–1.5 mm. This reflected radiation is collected by a phototransistor positioned beside the LED in the PPG probe, which is held into position by double-sided adhesive tape (Fig. 3). The signal is processed by the photoplethysmograph and displayed as a fluctuating voltage on a chart recorder. Whereas the recording displayed by the LDV apparatus can be damped so that output oscillations due to the heartbeat can be eliminated, this is not possible with the PPG instrument. Hence the results from the PPG appear as a series of pulses on which, at high gain, the dicrotic notch may be observed (Fig. 4).

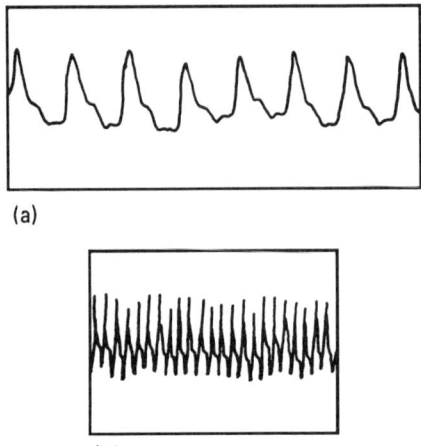

Figure 4 Typical pulsatile signal from the PPG device with the probe attached to the forefinger. The dicrotic notch is visible at both the fast (A−25 mm/sec) and the slow (B−5 mm/sec) chart recorder speeds.

APPLICATIONS

Noninvasive optical techniques for determining blood flow in cutaneous microcirculation are increasingly being utilized in a broad range of applications. The usefulness of LDV and PPG is limited by their inability to provide absolute values of blood flow, and as noted earlier, this information may be obtained by calibrating the instrument with the xenon washout technique. Numerous studies have demonstrated the remarkable versatility of these noninvasive optical techniques in both clinical and experimental settings. A brief review of these data is presented, not in an attempt to delineate all previous applications, but rather to demonstrate the varied and possible uses of LDV and PPG.

Clinical Studies

Both LDV and PPG have been successfully utilized as objective predictors of cutaneous flap viability. Larrabee et al. compared LDV to conventional fluorescein dye injection for assessing flap circulation and subsequent survival (13). The procedures were equal in ability when measurements were made intraoperatively, and the LDV readings obtained 24 hr following the surgery were excellent markers of flap viability, leading these authors to suggest that LDV is a practical alternative to fluorescein dye evaluation. An earlier review by Challoner demonstrated the usefulness of PPG in determining the viability of tubed pedicle flaps (12). Although most reconstructive surgeons continue to rely predominantly on clinical assessment of flap viability, noninvasive optical techniques offer significant information in helping to determine the final outcome of cutaneous flaps at risk.

Rosenberg et al. utilized LDV to assess the viability of cutaneous tissue subjected to thermal injury (14). Calibration studies were done using a model of controlled thermal injury in rats, and a good correlation between histological evaluation and LDV measured perfusion was obtained. They concluded that LDV was a useful and practical alternative for measuring capillary blood flow, especially with respect to the continuous, immediate nature of the monitoring and the noninvasiveness of the procedure.

The work of Kristensen et al. suggests that LDV may be of benefit in the assessment and classification of Raynaud's phenomenon (15). Measurements of digital blood flow in normal subjects and in scleroderma patients with Raynaud's phenomenon revealed that in the latter, fingertip blood flow reacted in the same way as normal skin without shunt vessels. During cooling, finger blood flow in patients with secondary Raynaud's phenomenon showed the same relative reduction as normal patients, but resting blood flow in these patients was significantly less and the rewarming period was markedly prolonged.

Cutaneous blood flow during pregnancy was investigated by Katz and Sokal utilizing PPG (16). Perfusion was found to increase significantly at the 16th gestational week and remain elevated until at least 1 week postpartum. The maximum elevation occurred between the 20th and 30th week of gestation and was 3--4 times the baseline values. They concluded that the results supported an inverse relationship between cutaneous blood flow during pregnancy and peripheral resistance.

Experimental Studies

The use of LDV and PPG to investigate alterations in cutaneous blood flow following local or systemic administration of vasoactive agents is a logical extension of these modalities. Hertzman and Randall pioneered the use of PPG and were among the first investigators to examine local vasodilation with this method (17). Recently LDV and PPG have been utilized in a systematic way to examine the local pharmacodynamic response to the topically applied vasodilator methyl nicotinate (18–21). This work enabled Guy et al. to assess the percutaneous absorption kinetics of this drug allowing measurement of penetration, residence, and elimination of methyl nicotinate from the local region of skin where it was applied. Several sets of experiments have allowed for the measurement of the initial onset of erythema (18) and the time course of the vasodilatory response at various anatomical sites (19). The dose-response behavior was generated over a nicotinate concentration range of 5–100 mM as well as for variations in the time of drug application and area of administration (21). This work demonstrates the versatility of both LDV and PPG in analyzing the pharmacodynamic profiles of topically vasoactive compounds, thus providing quantitative measurements of otherwise inaccessible aspects of in vivo transcutaneous pharmacokinetics. Preliminary studies by Wester et al. examining the local effects of topical minoxidil using LDV demonstrated measurable vasodilation following two daily applications of a 5% concentration to the scalps of balding male subjects (22). Vasodilation of the cutaneous microcirculation of the fingertips following the systemic administration of nitroglycerin sublingually was also elegantly demonstrated by Wester et al. utilizing LDV (23). Concurrent determination of plasma nitroglycerin levels showed excellent correlation between the LDV data and the plasma levels of the drug. This work provides a nice example of the usefulness of noninvasive optical techniques in determining the pharmacokinetic and pharmacodynamic profiles of systemically administered vasoactive compounds.

Irritation

Assessment of skin irritancy reactions by objective measurement of resultant erythema has been investigated by Nilsson et al. (24,25). The use of LDV to determine the degree of vasodilation following topical application of primary irritants offers a significant advantage over subjective visual scoring of erythema. This was shown for skin sites exposed to sodium lauryl sulfate, under occlusion, in concentrations ranging from 0.001% to 5%, graded visually and measured by LDV at 26, 48, and 72 hr after application. Although good correlation was observed, some data suggested that visual assessment was significantly less reliable than LDV. In a similar study assessing the irritancy of propylene glycol, a slight increase in blood flow was demonstrated with LDV on the volar forearm following occlusion, which was only detectable as a faint erythema by unaided visual observation (25).

Noninvasive optical techniques have been utilized to examine the intensity and time course of ultraviolet (UV) radiation—induced erythema. Initial work by Ramsay and Cripps, using PPG, measured cutaneous arteriolar dilatation following UV irradiation at either 250 or 300 nm, demonstrating identical responses at both wavelengths (26). Additional work confirmed these findings, suggesting that cutaneous blood flow changes induced by UV irradiation at both 250 and 300 nm are similar and may be mediated by identical mechanisms (27). Initial measurements of UV-induced erythema utilizing LDV were carried out by Stern et al. (1). UV irradiation was performed 24 hr prior to LDV monitoring of treatment and control sites, which revealed perfusion to be 2.7 times higher at irradiated sites. A reasonably linear correlation existed between the LDV determinations and the corresponding 133-Xenon washout flow measurements which were concurrently performed. Recently Drouard et al. utilized LDV to quantify UV-induced changes in microcirculatory flow and determine the potential use of this procedure for evaluating sunscreen efficacy in actual use situations and the effects of specific constituents in sunscreen formulations (18). Blood flow in the cutaneous microcirculation of the upper back was monitored following UV irradiation of untreated control skin and skin pretreated with sunscreen. LDV had the capability to objectively and quantitatively assess and monitor UV-induced erythema, which correlated well with visual assessment, and demonstrated the protective effects of the active sunscreen compound (2-ethylhexylcinnamate) without evidence of either enhancement or diminution of the protective effect of the cinnamate by 30 ppm of 5-methoxypsoralen. Overall, they concluded that the objective nature of the measurement and the noninvasive character of LDV offered considerable advantages over other currently available methods of determining UV induced changes in cutaneous microcirculatory blood flow.

CONCLUSIONS

The noninvasive optical techniques of PPG and LDV have been widely utilized for measurement of cutaneous microcirulatory blood flow. In most instances the results obtained have been reliable and easily reproducible, effectively fulfilling the requirements of each investigator. The ease of performing repeated or continuous measurements in a noninvasive manner makes these methods particularly well suited for clinical studies in reconstructive plastic surgery, peripheral vascular disease, and dermatology. The objective qualitative results yielded by these techniques makes them invaluable to any cutaneous investigator interested in alterations in microvascular blood flow. Of particular note is the proven usefulness of these modalities in assessing the pharmacodynamic response of topically applied medications. This aspect of LDV and PPG will find broad application in the fields of dermatopharmacology, cosmetic chemistry, and cutaneous physiology.

REFERENCES

1. Stern, M. D., Lappe, D. L., Bowen, P. D., Chimosky, J. E., Holloway, G. A., Keiser, H. R., and Bowman, R. L. Continuous measurement of tissue blood flow by laser-Doppler spectroscopy. *Am. J. Physiol.* 232:441–448 (1977).

2. Tur, E., Tur, M., Maibach, H. I., and Guy, R. H. Basal perfusion of the cutaneous microcirculation: Measurements as a function of anatomic position. *J. Invest. Dermatol.* 81: 442–446 (1983).

3. Guy, R. H., Tur, E., and Maibach, H. I. Optical techniques for monitoring cutaneous microcirculation. *Int. J. Dermatol.* 24:88–94 (1985).

4. Stern, M. D. In vivo evaluation of microcirculation by coherent light scattering. *Nature* 254:56–58 (1975).

5. Holloway, G. A., and Watkins, D. W. Laser Doppler measurement of cutaneous blood flow. *J. Invest. Dermatol.* 69: 306–309 (1977).

6. Watkins, D. W., and Holloway, G. A. An instrument to measure cutaneous blood flow using the Doppler shift of laser light. *IEEE Trans. Biomed. Eng.* BME 25:28–33 (1978).

7. Bonner, R. F., Clem, T. R., Bowen, P. D., and Bowman, R. L. Laser Doppler continuous real-time monitor of pulsatile and mean blood flow in tissue microcirculation. In: *Scattering Techniques Applied to Supramolecular and Non-equilibrium*

Systems, NATO, ASI Series B. Vol. 73 (S. Chen, B. Chen, and R. Nossal, eds.). Plenum Press, New York, pp. 685–702 (1981).

8. Nilsson, G. E., Tenland, T., and Oberg, P. A. A new instrument for continuous measurement of tissue blood flow by light beating spectroscopy. *IEEE Trans. Biomed. Eng.* BME 27:12–19 (1980).

9. Nilsson, G. E., Tenland, T., and Oberg, P. A. Evaluation of a laser Doppler flowmeter for measurement of tissue blood flow. *IEEE Trans. Biomed. Eng.* BME 27:597–604 (1980).

10. Englehart, M., and Kristensen, J. K. Evaluation of cutaneous blood flow responses by 133Xe washout and a laser Doppler flowmeter. *J. Invest. Dermatol.* 80:12–15 (1983).

11. Ware, B. R. Laser Doppler velocimetry. *Am. Lab.* 17–26 (1981).

12. Challoner, A. V. Photoelectric plethysmography for estimating cutaneous blood flow. In: *Non-invasive Physiological Measurements.* Volume 1 (P. Rolfe, ed.). Academic Press, London, pp. 125–151 (1979).

13. Larrabee, W., Holloway, G., and Sutton, D. A comparison of laser Doppler velocimetry and fluorescent dye in the prediction of flap viability. Abstract presented at the American Academy of Facial Plastic and Reconstructive Surgery, New Orleans (1982).

14. Rosenberg, L., Molcho, J., and Dotan, Y. Use of Doppler effect in visible laser light to assess tissue viability by capillary blood flow. *Ann. Plastic Surg.* 8:206–212 (1982).

15. Kristensen, J. K., Engelhart, M., and Nielsen, T. Laser-Doppler measurement of digital blood flow regulation in normals and in patients with Raynaud's phenomenon. *Acta Dermatol. Venereol.* 63:43–47 (1983).

16. Katz, M., and Sokal, M. M. Skin perfusion in pregnancy. *Am. J. Obstet. Gynecol.* 137:30–33 (1980).

17. Hertzman, A. B., and Randall, W. C. Regional differences in the basal and maximal rates of blood flow in the skin. *J. Appl. Physiol.* 1:234–241 (1948).

18. Guy, R. H., Wester, R. C., Tur, E., and Maibach, H. I. Noninvasive assessment of the percutaneous absorption of methyl nicotinate in humans. *J. Pharm Sci.* 72:1077–1079 (1983).

19. Tur, E., Guy, R. H., Tur, M., and Maibach, H. I. Noninvasive assessment of local nicotinate pharmacodynamics by photoplethysmography. *J. Invest. Dermatol.* 80:499–503 (1983).

20. Amantea, M., Tur, E., Maibach, H. I., and Guy, R. H. Preliminary skin blood flow measurements appear unsuccessful for assessing corticosteroid effect. *Arch. Dermatol. Res.* 275:419–420 (1983).

21. Guy, R. H., Tur, E., Bugatto, B., Gaebel, C., Sheiner, L. B., and Maibach, H. I. Pharmacodynamic measurement of methyl nicotinate percutaneous absorption. *Pharm. Res.* 1:76–81 (1984).

22. Wester, R. C., Maibach, H. I., Guy, R. H., and Novak, E. Minoxidil stimulates cutaneous blood flow in human balding scalps: pharmacodynamics measured by laser Doppler velocimetry and photopulse plethysmography. *J. Invest. Dermatol.* 82:515–517 (1984).

23. Wester, R. C., Maibach, H. I., and Guy, R. H. Nitroglycerin pharmacodynamics monitored noninvasively by laser Doppler velocimetry. Abstract presented at the 84th Annual Meeting of the American Society for Clinical Pharacology and Therapeutics, San Diego (1983).

24. Nilsson, G. E., Otto, U., and Wahlberg, J. E. Assessment of skin irritancy in man by laser Doppler flowmetry. *Contact Dermatitis* 8:401–406 (1982).

25. Wahlberg, J. E., and Nilsson, G. E. Skin irritancy from propylene glycol. *Acta Dermatol. Venereol.* 64:286–290 (1984).

26. Ramsay, C. A., and Cripps, D. J. Cutaneous arteriolar dilatation elicited by ultraviolet irradiation. *J. Invest. Dermatol.* 54:332–337 (1970).

27. Ramsay, C. A., and Challoner, A. V. L. Vascular changes in human skin after ultraviolet irradiation. *Br. J. Dermatol.* 94:487–493 (1976).

28. Drouard, V., Wilson, D. R., Maibach, H. I., and Guy, R. H. Quantitative assessment of UV-induced changes in microcirculatory flow by laser Doppler velocimetry. *J. Invest. Dermatol.* 83:188–192 (1984).

7
Skin Impedance Measurement

WILLIAM I. ARCHER*, R. KOHLI[†], and J. M. C. ROBERTS[‡]
Johnson Wax Ltd., Egham, England

THOMAS S. SPENCER *Cygnus Research Corporation, Redwood City, California*

INTRODUCTION

The aim of this chapter is to introduce the reader to the use of alternating-current (a.c.) impedance as a technique for assessing skin properties. This will be achieved by first briefly reviewing the published literature and then discussing the basic theories involved and the instrumentation available. Specific experimental details, problems encountered, and possible solutions will then be presented followed by a discussion of some typical results.

Present affiliations:
*S. C. Johnson & Son, Inc., Racine, Wisconsin.
[†]Colgate Palmolive, Herstal Milmort, Belgium.
[‡]S. C. Johnson & Son, Inc., Racine, Wisconsin.

HISTORY OF ELECTRICAL CHARACTERIZATION OF THE SKIN

Various electrical measurements have been made on the human body for a number of years (1–13). However, there has been a perceptible growth in interest in recent years, partly due to developments in electronics and available apparatus, but also due to an increasing interest in the structure and function of the skin (14–25).

An additional factor cited as favoring the use of electrical techniques to characterize the skin is that electrical changes with function are much greater as a percentage of the baseline than is the case with other commonly employed methods, such as measurement of thermal or mechanical properties (32).

The electrical studies previously undertaken on the skin have shown a wide variation in the sophistication of the equipment and techniques employed, the objectives of the studies, and the complexity of models used to aid in interpretation of the data (1–30).

Direct current studies (and low-frequency a.c. studies) often given rise to polarization effects which lead to incorrect values for the electrical parameters required. It is thus preferable to conduct measurements using a small-amplitude alternating signal at frequencies above those at which polarization effects can occur. This is the major reason for the increasing use of a.c. impedance as a method for investigating skin.

A more detailed review of skin impedance studies is given later; however, it is beyond the scope of this chapter to give detailed coverage of other electrical measurement techniques. The reader is therefore referred to a number of comprehensive reviews already available in the literature (26–32).

SUMMARY OF BASIC a.c. IMPEDANCE THEORY

In essence, a.c. impedance studies of any system involve the application of a potential (or current), which is time-dependent in a sinusoidal manner, i.e., $\Delta E \sin \omega t$ (where ΔE is small—typically <10 mV), between the two electrodes situated on either side of the test substrate. As a consequence of the application of this perturbing voltage, a sinusoidal current (or voltage) will flow, having the form $\Delta i \sin (\omega t + \theta)$. It should be noted that ω equals $2\pi f$, where f is the sinusoidal frequency in Hz.

Thus, one can define an impedance z as having a magnitude given by:

$$|z| = \frac{\Delta E}{\Delta i} \tag{1}$$

and a phase angle θ, which corresponds to the phase difference between the applied sinusoidal potential and the resultant sinusoidal current. An impedance is, therefore, a vector quantity, since it has both magnitude and direction. It is convenient to represent such two-component vectors as a point in a plane (Fig. 1), where it can be characterized by $|Z|$ and θ or by its real and imaginary components, Z' and Z" projected on the X and Y axes.

When the test cell is replaced by a pure resistance of magnitude R ohms, we find that Z = R and θ = 0. Thus, a pure resistance is represented by a point on the X axis for any sinusoidal frequency (Fig. 2a). When a pure capacitance C (farads) is substituted for the cell, the situation becomes somewhat more complicated. In this case, θ = 90°, but Z is frequency-dependent through the relationship $Z = 1/\omega C$. Therefore, as the

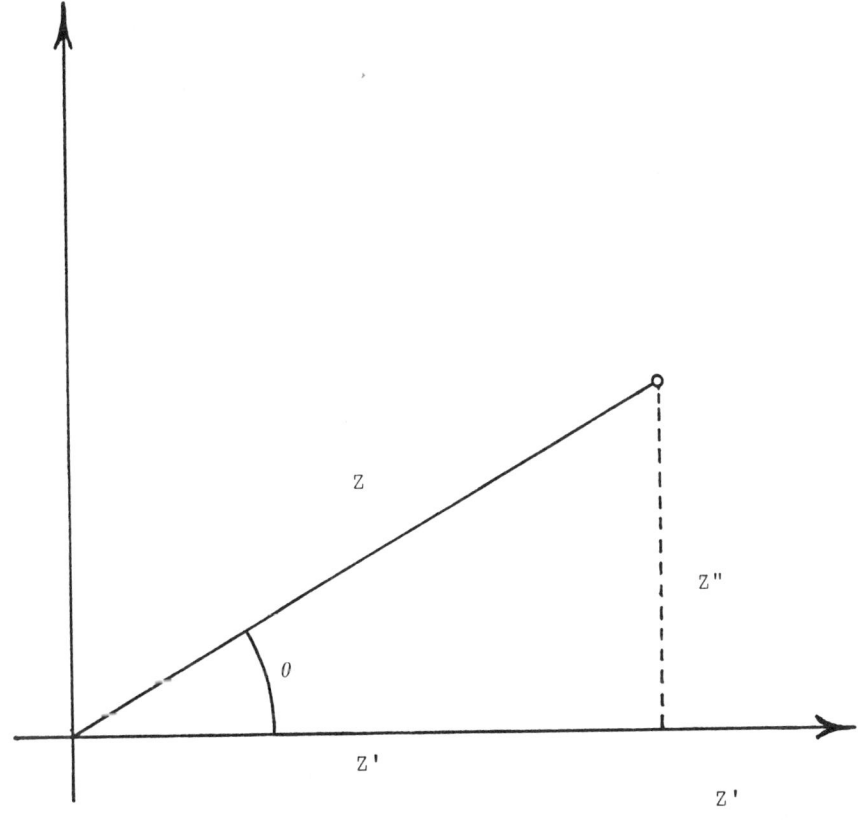

Figure 1 Complex plane representation of impedance.

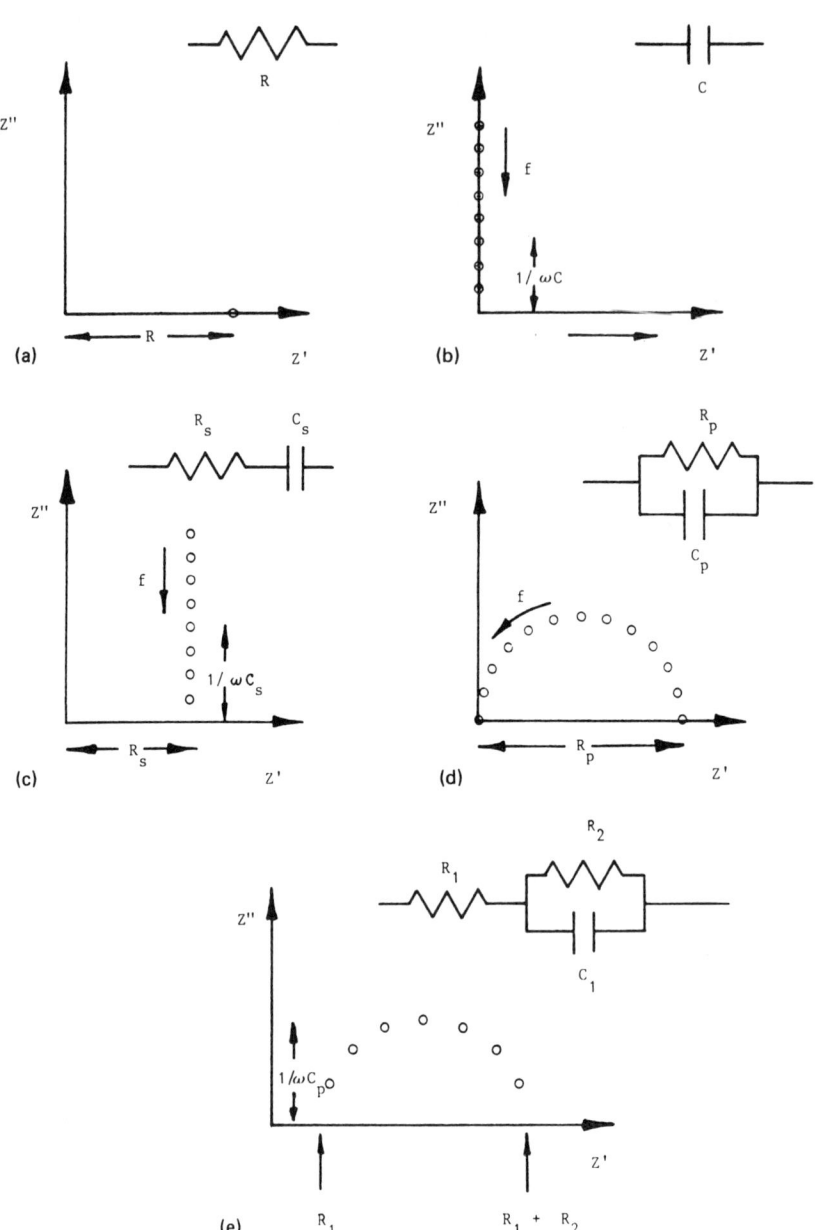

Figure 2 Complex plane impedance spectra with their associated equivalent circuits.

frequency of the sinusoidal potential is varied, the representative point in the complex plane also varies as in Figure 2b. Figures 2a and 2b, are the simplest forms of complex-plane impedance spectra, which are representations of the impedance being measured as a function of frequency. Plots of this kind are also referred to as Argand diagrams, Nyquist plots, or Cole-Cole plots.

In the majority of cases, the test cell is better represented by a more elaborate network of resistances and capacitances, the so-called equivalent circuit. These show a more complex behavior in the impedance plane. For example, a resistance and capacitance in series gives the impedance spectrum shown in Figure 2c. In this case,

$$Z' = R_s \text{ and } Z'' = \frac{1}{\omega C_s}$$

and since

$$Z^2 = Z'^2 + Z''^2$$

then

$$Z^2 = R_s^2 + (1/\omega C_s)^2 \tag{2}$$

and θ can take any value between $0°$ and $90°$, depending on the measurement frequency. From this, one can derive an equation for the impedance:

$$Z = R_s - \frac{j}{\omega C_s} \tag{3}$$

where $j = \sqrt{-1}$.

If, instead of a series combination of resistance and capacitance, we have a parallel circuit, the impedance spectrum is quite different (Fig. 2d), and in this case, the impedance is given by

$$Z = [(1/R_p) + j \omega C_p]^{-1} \tag{4}$$

When using a parallel combination of elements, the capacitance is additive, which can be advantageous. It is, therefore, sometimes more useful to plot the admittance Y, given by

$$Y = \frac{1}{Z} = \frac{1}{R_p} + j\omega C_p \tag{5}$$

in the same manner described above. A valuable set of equations used to interconvert admittance and impedance, and hence showing the relationships between R_p and R_s and C_p and C_s, can be found in a review by Archer and Armstrong (35). Readers who desire a more rigorous mathematical treatment of the electrical impedance behavior of biological materials are referred to two publications by Salter (32,33).

Given an impedance (or admittance) spectrum, the components of an equivalent circuit of resistances and capacitances can be calculated (e.g., Fig. 2e). With measurements on experimental systems, such as the skin, it is more usual for the investigator to measure the impedance of the test cell and subsequently to try to find the appropriate equivalent circuit and then identify the physical significance of the various components. This may be achieved by comparison of the results with theoretical models (14,21,22,33,34) or by devising an equivalent circuit that duplicates the experimental behavior.

APPLICATION OF IMPEDANCE TECHNIQUES TO THE SKIN

The properties of cells or tissue, such as the skin, are related to changes in the metabolic, physiological, and psychological state of the parent organism. The skin, being on the external surface of the body, is also susceptible to environmental conditions, particularly temperature and humidity. Thus, the biological, physical, and, hence, biophysical properties of the skin are affected by changes in both the external and internal conditions. This is especially true of the electrical properties of the skin.

The value of determining the electrical characteristics, such as the impedance of living tissues, in particular skin, has long been recognized, as such measurements may potentially provide considerable information relating to the substrate under investigation.

Equivalent Circuits and Anticipated Behavior

For the purposes of electrical characterization, the skin can be divided into two regions: the highly conductive inner tissue and the more resistive stratum corneum.

A typical complex plane impedance diagram obtained from measurements of the skin takes the form of a circular arc, with its center depressed below the real axis (Fig. 3) and offset from the origin. Here R_{it} is the "inner tissue resistance," i.e., the resistance of the sub-stratum corneum tissue, and R_{sc} is the stratum corneum resistance.

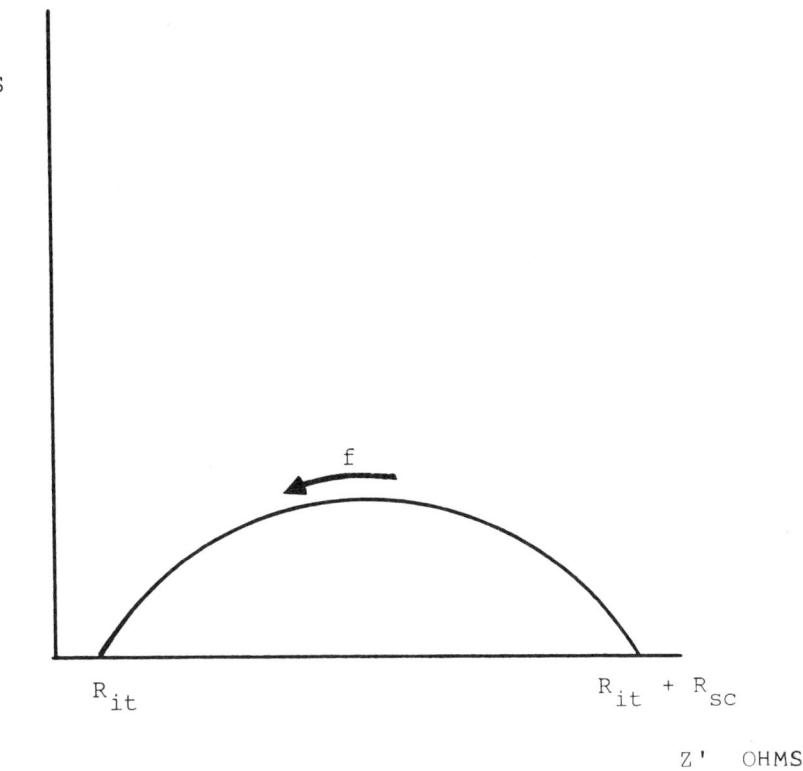

Figure 3 Typical complex plane impedance spectrum obtained for skin.

Figure 4 shows the equivalent circuit that is commonly accepted as modeling the impedance behavior of the skin; that is, the impedance response of this circuit is of the type shown in Figure 3. In this circuit, the boxed element is known as a "polarization impedance." Such elements are generally included in equivalent circuits in order to accommodate a distribution of relaxation times that occurs in the material under consideration. This distribution of relaxation times, which is implied whenever a "depressed semicircle" is obtained in the complex plane, in the case of skin may arise from within each individual cell membrane or as a result

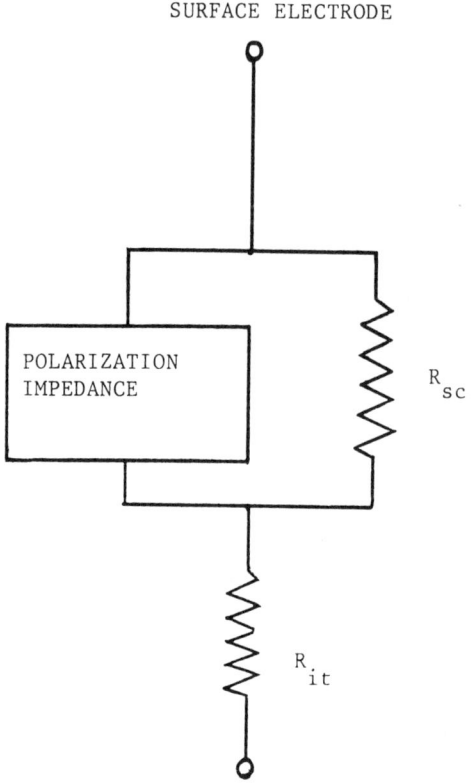

Figure 4 Equivalent circuit for skin impedance. (This represents the situation for one surface electrode and one subdermal electrode.)

of the gross inhomogeneity within the skin, or from a combination of the two. Salter (32) gives a comprehensive, detailed review and discussion of the concept of polarization impedance for the reader who requires further information in this respect.

What Kind of Information May Be Obtained?

The majority of recent electrical studies on the skin have concentrated on the various aspects of skin hydration or moisturization

(15–20, 23–25, 40, 41). Leveque and de Rigal (26) have recently, in a review, identified at least three possible conduction mechanisms that may occur within the skin and be affected by the water content of the stratum corneum.

1. The keratin chains in the straum corneum, which have a dipole moment, may be made more mobile by the plasticizing effect of water.
2. Ions in the intercellular spaces can react to the application of any electrical field and may, to a certain extent, move with it if they have a sufficiently high mobility. The mobility of these ions will be dependent upon, among other factors, the viscosity of the medium, which, in turn, depends upon the water content.
3. The water molecules present in the skin may form a continuous network of hydrogen bonds, allowing the exchange of a proton between adjacent molecules.

It is thus reasonable to expect that changes in skin hydration/moisturization will cause changes in the electrical properties of the skin and that these changes could be detected and quantified by impedance studies.

Since the hydration level of the skin, particularly the stratum corneum, is of great interest to the cosmetic scientist and dermatologist alike, as an indication of cosmetic product efficacy and skin condition on a clinical basis, skin impedance studies potentially offer an extremely valuable method for studying such parameters.

On this basis, any topical application to the skin that results in changes in the electrical properties of the skin may be studied by impedance methods. This broadens the field of study from moisturization alone to encompass oiling, irritation, transepidermal drug delivery, and defatting, to name but a few areas that merit investigation in this way.

EXPERIMENTAL PROCEDURES

Instrumentation

The simplest experimental arrangement for measuring the impedance of a biological system involves the use of an a.c. bridge. Here, the test cell, which is in one arm of the bridge, is balanced against decade capacitance and resistance boxes in another arm. If, at the balance point of the bridge, the components of the impedance are R_S and C_S, then $Z' = R_S$ and $Z'' = 1/\omega C_S$. However, it is laborious to obtain measurements over a frequency range in order to construct an impedance spectrum since a balance point takes several minutes at each frequency.

The direct measurement of the in-phase and quadrature components of potential and current using phase-sensitive detectors (PSD) gives considerable improvement. This gives a direct-current output that is related to both the amplitude and the phase of the sine-wave input. By suitable phase shifts, one can obtain the real and imaginary parts of the impedance. Thus, in this case, no balancing is involved, but to obtain an impedance spectrum with approximately five points for each decade in frequency down to 1 Hz may take 2 hr or more. Below this frequency, the experimental time increases rapidly because of the time needed for accurate averaging by the PSD. Furthermore, the data obtained generally need to be processed numerically before the impedance can be obtained.

A more sophisticated experimental arrangement is based on the Solartron 1170 and 1250 series Frequency Response Analysers (FRA) (36). This system removes a great deal of the tedium involved in simple impedance measurements and allows a large frequency range to be covered more quickly, and usually more accurately. Armstrong et al. have covered the use of an 1170-based system in detail (37) and the majority of the principles also apply to the more recently available 1250 series machines. The 1250 FRA's are more sophisticated and quicker, and they also offer the capability of simultaneous measurement of more than one test site with one instrument. Essentially, the FRA consists of a programmable generator that provides the perturbing sinusoidal signal, a correlator to analyze the response of the system, and a display to present the results.

The FRA analyzes the response of the system by a correlation process to determine Δi and θ. The correlation process has the advantages of rejecting all harmonics present in the system output and minimizing the effects of random noise. A single measurement at a particular frequency can be made by programming the generator with the required frequency and signal amplitude. More usually, however, the generator is programmed to sweep through a large frequency range by choosing suitable values of the maximum frequency, the minimum frequency, and the number of points per decade at which measurements are to be taken. The instrument will then take measurements sequentially, in either direction, at equally spaced intervals, on either a logarithmic or a linear scale, over the required range. The response is given once a measurement has been completed and can be displayed in any of three possible notations: (1) amplitude (A) and phase angle (θ)

Skin Impedance Measurement

relative to the input signal, (2) log A and (θ), or (3) the real and imaginary parts of the impedance. It is now common practice to have the FRA interfaced to a computer so that a real-time plot of the measured data is obtained on the display of the computer. The data are then stored on an appropriate mass storage device, from where they can be recalled for further manipulations. Using Solartron Frequency Response Analyzers, measurements can be performed very quickly, with a complex plane impedance spectrum over the frequency range of interest being obtained in a matter of seconds. (At frequencies below 1 Hz, the time for a single measurement is dependent on the period of the signal and gets progressively slower as the frequency gets lower.)

The methods of impedance determination discussed up to this point can all be considered direct methods. Indirect determination of the transfer function has also been performed by Laplace and Fast Fourier transforms, and a number of systems are now commercially available, making use of computerized signal processing techniques (38,39).

A number of techniques have also been described recently, which cover impedance measurement in the very-high-frequency range, 10 MHz to 12 GHz (26,40,41).

Electrodes

Once it has been decided to undertake skin impedance measurements, a choice must be made as to the type, number, and arrangement of the electrodes to be used. In deciding which type of electrodes to use, a number of factors should be kept in mind.

It is likely that a "contact impedance" will exist between the electrodes and the skin. This then results in a pressure dependence of the measured impedance. The extent of this impedance is critically dependent on the choice of electrode. The use of a contact gel between the electrodes and the skin will minimize the effects of any contact impedance; however, careful consideration must also be given to the choice of gel, as the gel itself may modify the skin or introduce anomalous impedance elements into the system.

Application of electrodes to the skin surface is likely to result in a buildup of moisture below the electrodes due to transepidermal water loss effects. Various attempts to counteract this problem have been made, ranging from open grid electrodes (20), which minimize the moisture buildup, to use of a solution contact designed to take into account the temperature and relative humidity of the environment, so that there is no net moisture exchange between the skin and the solution (15, 23).

In our opinion, the choice of electrodes involves a compromise between the above factors, which often have conflicting demands, and ease of use. If the behavior of the electrodes and their influence on the measurements have been studied in detail and are well understood, one can use various arrangements. In fact, we prefer to use commercially available silver/silver chloride electrodes (In Vivo Metric Systems, Healdsburg, CA) with a salt-free electrode contact gel (Spectra 360, In Vivo Metric Systems).

The number of electrodes employed has also varied, from four to two, involving geometries ranging from linear arrangements to concentric electrodes. A number of specialized electrode assemblies have been produced (16, 41) in attempts to produce electrode separations comparable to the stratum corneum thickness. However, in doing this, one greatly increases the risk of short-circuiting by surface conduction. Lykken (42) reviewed the use of different electrode materials, and additional information relating to electrode selection can be found in a number of reviews (26–32). In our laboratories, we use a two-electrode symmetrical arrangement, which offers advantages both in ease of use and in ease of interpretation of the results.

Experimental Conditions

Our experience in this area has indicated that the experimental conditions must be carefully controlled. For this reason we conduct our measurements in an environmentally controlled room where the subject is equilibrated and relaxed prior to measurement.

It is important to control, as far as possible, any disturbances that may affect the autonomic response of the subject. These include interruptions, talking to the subject, and any other spurious occurrences. We have also found that it is possible, to a certain extent, to influence the results as they are produced, if the subject can see the results displayed while the measurement is in progress.

Obviously, many factors, including exercise, diet, intake of stimulants, phychological state, and state of health, affect the results obtained. It is impossible to control all of these at any one time. To overcome this, we have used the subject as his own internal control, by using two identical pairs of electrodes, one as a test site and one as a control site, which should show any trends in the baseline values.

The majority of our measurements have been made using Solartron Frequency Response Analysers and the previously described Ag/AgCl electrodes. Figure 5 shows a typical experimental arrangement using an 1174 Frequency Response Analyser. The current measuring resistance is used to measure the current response (as a voltage drop across a resistance) and its value should be chosen to be of the same order of magnitude as the real impedance component at the particular frequency of study. This maximizes the accuracy of measurement and minimizes scatter in the data.

Up to the present time, we have measured control and test sites alternately by switching; however, we are currently developing software to enable us to use the four channel analyzer facilities of the 1254 FRA in order to measure test and control sites simultaneously.

Our normal frequency range of measurement is from 1 to 500 Hz, with a generator output of 100 mV. At these frequencies, capacitative artifacts from lengthy leads and bad cell design are minimal. However, it is important to consider these factors when making measurements above about 10 kHz.

The input impedance of the FRA is 1 Mohm, which tends to be comparable to or less than the measured impedance. It is, therefore, necessary to increase the input impedance of the system. This can be achieved in a number of ways, two of which are favored by us. Simply using 10× matched oscilloscope probes increases the effective input impedance to 10 Mohms; alternatively, the use of the Solartron 1186 or 1286 electrochemical interfaces increases the input impedance to around 10 Gohms.

Data Analysis

The experimental arrangement shown diagramatically in Figure 5 gives a real-time display of the complex plane impedance diagram. The data are then stored on a flexible disk and nonlinear least-squares routines are employed to give a curve fit to the data. Figure 6 shows a typical plot obtained for untreated skin. The continuous line represents the curve fit, and the values for R_{it} and R_{sc} can be obtained from the intercepts on the X axis. Plots of R_{sc} vs. time can then be made, along with other plots as described earlier.

Single Point vs. Frequency Spectrum

Since the measured phase angle is dependent not only on the measurement frequency, but also on a number of other parameters

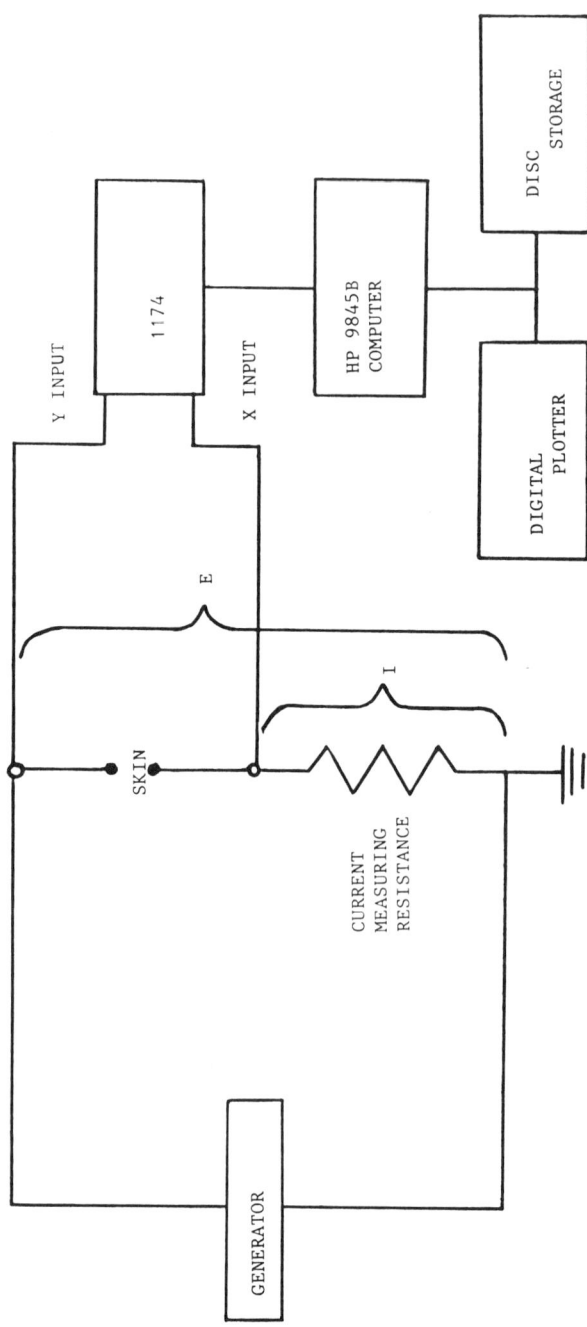

Figure 5 Typical experimental arrangement for skin impedance studies.

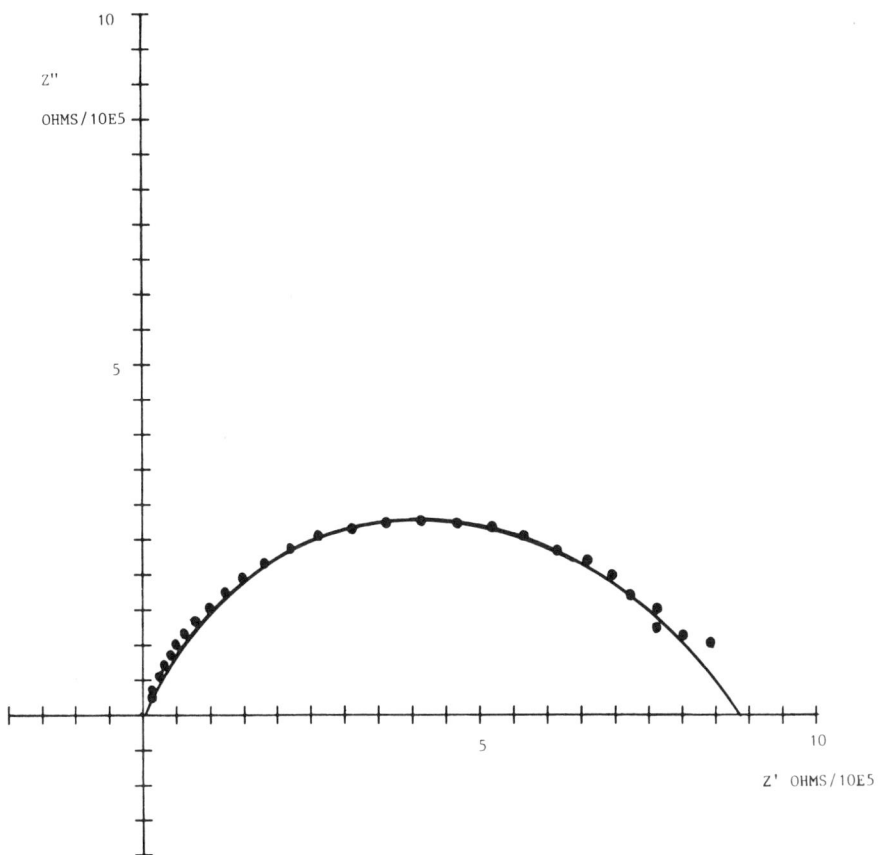

Figure 6 Typical impedance plot for untreated skin.

(see Ref. 32), it is essential to conduct measurements such that the full (or nearly full) complex plane impedance diagram is obtained. The use of single-point measurements, while perhaps rapid, is questionable in view of the fact that any trends seen cannot be related to a change in a single parameter and may, in fact, give rise to erroneous interpretation of results.

TYPICAL RESULTS

In this section, we shall use some of our own results to illustrate a number of important points. All of the results shown were obtained with the experimental arrangement of Figure 5 and are presented as R_{SC} vs. time plots.

Time Dependence and Intersubject Variability

Figure 7 shows the time dependence of the computed stratum corneum resistance, measurements being taken approximately every

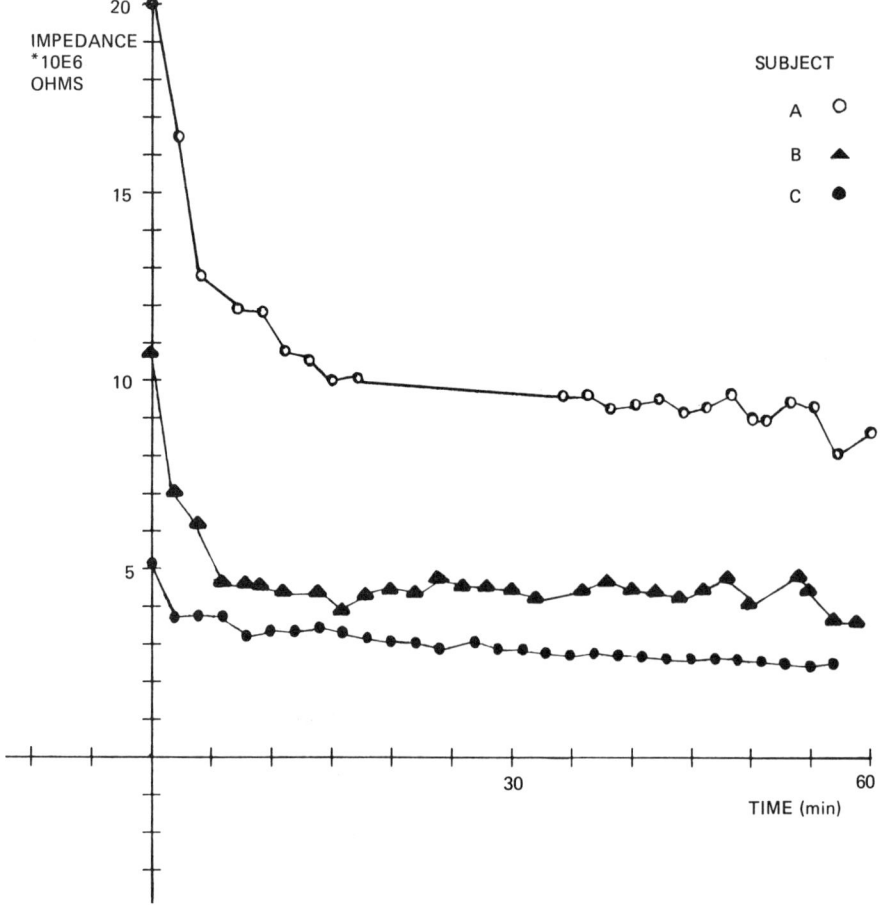

Figure 7 Variation of impedance with time.

2 min after application of the electrodes. Despite the fact that the subjects were "equilibrated" for 30 min prior to measurement, it can be seen that there is an initial decay in the R_{sc} value and only after some 10 or more minutes is a reasonably steady state achieved.

We attribute the initial rapid decay in the R_{sc} value to some function of the electrode/skin equilibration rather than to a buildup of moisture in the straum corneum due to transepidermal water loss.

It is also interesting to note the wide intersubject variability for the three subjects shown. Even though the subjects were all Caucasians in the age range 21−25, there is a considerable spread in the measured impedance.

Day-to-Day Variation

Measurements conducted on the same group of subjects over a period of days shows that there is a wide day-to-day variability within any one subject and that there is no clear trend affecting all subjects.

These observations can be accounted for by changes in the physiological and psychological state of the subjects. If the changes were due simply to environmental changes, it is likely that a consistency would be observed throughout the group of subjects.

Site-to-Site Variation

We have observed a considerable variation in impedance from site to site. This is to be anticipated when one considers the differences in stratum corneum thickness, sweat pore density, and so on between sites. However, even in an area such as the volar surface of the forearm, we have observed a degree of variability from both left to right and between wrist and elbows. Figure 8 gives a typical example of the results obtained at the wrist and elbow of the same arm, measured over the same period. These variations can be attributed to variations in stratum corneum thickness or other localized inhomogeneities.

Electrical Signal Path

We have conducted a number of studies involving variations of electrode separation, stratum corneum stripping, and topical applications in efforts to elucidate the signal path through the skin,

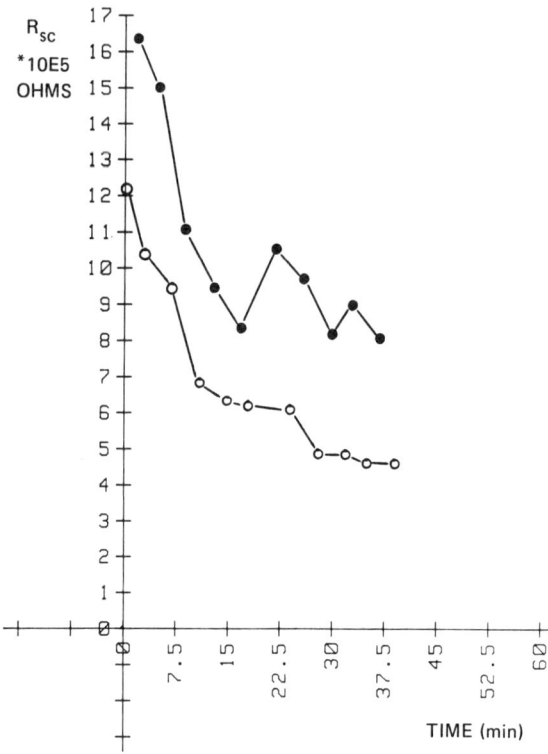

Figure 8 Variation of impedance between two forearm sites.

which is appropriate to our experimental arrangement. Figure 9 shows the signal paths we consider appropriate in our case for untreated skin. Obviously the diagram is only schematic and a great deal of further work and modeling are required to accurately reproduce the exact distribution of the electrical field in the skin.

SUMMARY

In this chapter we have aimed to provide the reader with sufficient theoretical and practical information to make skin impedance studies. These studies may offer a number of advantages over

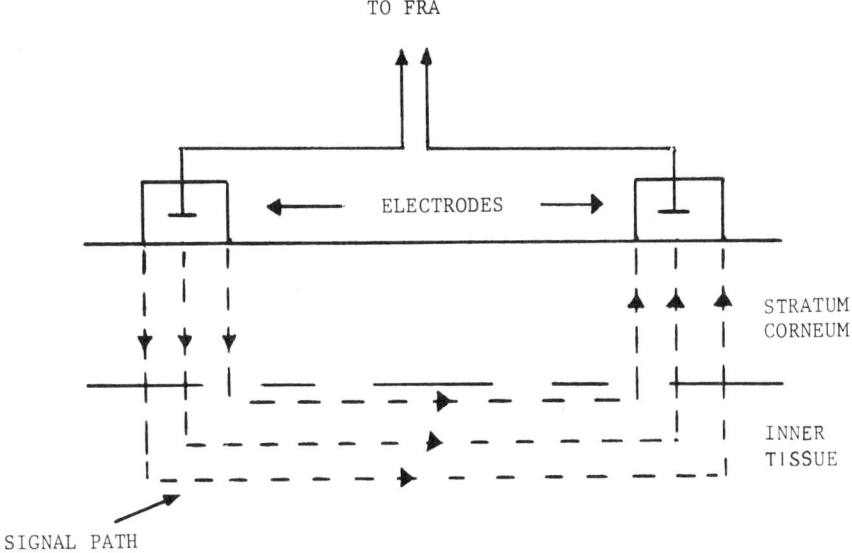

Figure 9 Schematic representation of the signal path through the skin.

other methods for skin assessment, but particular care is required in the measurement of controls such that the problem of high baseline variability can be overcome. Provided that this difficulty can be overcome experimentally, a potentially large amount of information is available from skin impedance studies.

REFERENCES

1. Fere, C. Note sur des modifications de la résistance électrique sous l'influence des excitations sensorielles et des émotions. C. R. Soc. Biol. Paris 8 (ser 5).217−219 (1888).

2. Vigouroux, R. De la résistance électrique comme signe clinique. Prog. Med. 3:87−89 (1888).

3. Gildermeister, M., and Kaufhold, R. Uber das elektrische Leitungsvermogen der uberlebenden menschlichen Haut. Pfluger's Arch. ges. Physiol. 179:154−158 (1920).

4. Gildermeister, M. Uber elektrishen widerstand, Kapazitat und Polarisation der Haut. II Mitteilung Menschliche Haut. *Pfluger's Arch. ges. Physiol.* 219:89–110 (1928).

5. Brazier, M. A. B. A method for the investigation of the impedance of the human body to an alternating current. *J. Inst. Elect. Eng.* 73:203 (1933).

6. Cole, K. S., and Curtis, H. J. Electric impedance of nerve and muscle. *Cold Spring Harb. Symp. Quant. Biol.* 4:73–89 (1936).

7. Thomas, P. E., and Korr, I. M. Relationship between sweat gland activity and electrical resistance of skin. *J. Appl. Physiol.* 10:505–510 (1957).

8. Lawler, J. C., et al. Electrical characteristics of the skin. *Invest. Dermatol.* 34:301–308 (1960).

9. Johnson, L. C. Racial differences in skin resistance. *Science* 134:766–767 (1963).

10. Tregear, R. T. Interpretation of skin impedance measurements. *Nature* 205:600–601 (1965).

11. Edelberg, R. Biopotentials from the skin surface: The hydration effect. *Ann. Acad. Sci.* 148:252–262 (1968).

12. Lykken, D. T., et al. Some properties of skin conductance and potential. *Psychophysiology* 5:253–268 (1968).

13. Allenby, A. C., et al. The effect of heat, pH and organic solvents on the electrical impedance and permeability of excised human skin. *Br. J. Dermatol.* 81 (Suppl 4):31–39 (1969).

14. Khalafalla, A. S., et al. An electrical model to simulate skin dielectric dispersion. *Computers Biomed Res.* 4:359–373 (1971).

15. Clar, E. J., et al. Skin impedance and moisturisation. *J. Soc. Cosmetic Chem.* 26:337–353 (1975).

16. Campbell, S. D., et al. Hydration characteristics and electrical resistivity of stratum corneum using a non-invasive four-point electrode method (in vitro). *J. Invest. Dermatol.* 69:290–295 (1977).

17. Jackson, R. J., et al. The measurement of the moisture content of the skin by a novel impedance technique. Proceedings ISCC Meeting, Venice (1980), pp. 667–676.

18. Tagami, H., et al. Evaluation of the skin surface hydration in vivo by electrical measurement. *J. Invest. Dermatol* 75: 500–507 (1980).
19. Serban, G. P., et al. In vivo evaluation of skin lotions by electrical capacitance. I. The effect of several lotions on the progression of damage and healing after repeated insult with sodium lauryl sulphate. *J. Soc. Cosmetic Chem.* 32: 407–419 (1981).
20. Serban, G. P., et al. In vivo evaluation of skin lotions by electrical capacitance. II. Evaluation of moisturized skin using an improved dry electrode. *J. Soc. Cosmetic Chem.* 32:421–435 (1981).
21. Yamamoto, T., and Yamamoto, Y. Non linear electrical properties of the skin in the low frequency range. *Med. Biol. Eng. Comput.* 19:302–310 (1981).
22. Poon, C. S., and Choy, T. T. C. Frequency dispersions of human skin dielectrics. *Biophys. J.* 34:135–147 (1981).
23. Clar, E. J., et al. Study of skin horny layer hydration and restoration by impedance measurement. *Cosmetics Toiletries* 94:33–40 (1982).
24. Tagami, H., et al. Water sorption-desorption test of the skin in vivo for functional assessment of the stratum corneum. *J. Invest. Dermatol.* 78:425–428 (1982).
25. Salter, D. C. Water absorption by the straum corneum studied using electrical impedance measurements. Presented at the 4th International Symposium on Bioengineering and the Skin, Besançon, France, September 1983.
26. Leveque, J. L., and de Rigal, J. Impedance methods for studying skin moisturisation. *J. Soc. Cosmet. Chem.* 34: 419 (1983).
27. Tregear, R. T. *Physical Functions of Skin.* Academic Press, London, New York (1966).
28. Edelberg, R. Electrical properties of skin. In: *Biophysical Properties of the Skin* (H. R. Elden, ed.). Wiley Interscience, New York (1971).
29. Fowles, D. C. Mechanisms of electrodermal activity. In: *Bio-electric Recording Techniques* (R. F. Thompson and M. Patterson, eds.). Academic Press, London, New York (1974).

30. Edelberg, R. Relation of electrical properties of skin to structure and physiologic state. *J. Invest. Dermatol.* 69: 324–327 (1977).
31. Millington, P. F., and Wilkinson, R. *Skin.* Cambridge University Press, New York, pp. 127–142 (1983).
32. Salter, D. C. Quantifying skin disease and healing in vivo using electrical impedance measurements. In: *Noninvasive Physiological Measurements*, Vol. 1 (P. Rolfe, ed.). Academic Press, London (1979).
33. Salter, D. C. Alternating current electrical properties of human skin measured in vivo. In: *Bioengineering and the Skin* (R. Marks and P. A. Payne, eds.). MTP Press, Lancaster, England, and Boston (1981).
34. Salter, D. C. The linear electrical properties of abnormal skin. *Acta Pharm. Suecia.* 20:63–65 (1983).
35. Archer, W. I., and Armstrong, R. D. The application of a.c. impedance methods to solid electrolytes. Specialist Periodical Reports, The Chemical Society, London. *Electrochemistry* 7:157 (1980).
36. Solartron Electronics Group Ltd., Farnborough, Hampshire, England.
37. Armstrong, R. D., et al. A method for automatic impedance measurement and analysis. *J. Electroanal. Chem. Interfacial Electrochem.* 77:287 (1977).
38. Wavetek Rockland Scientific, Inc., Rockleigh, NJ.
39. EG + G. Princeton Applied Research, Princeton, NJ.
40. Clar, E. J., and Sturelle, C. Etude des constituents du deme pour spectroscopic temparelle. *Parfums. Cosmet. Aromes.* 4.55 00 (1975).
41. Jacques, S. L., et al. Water content in stratum corneum measured by a focussed probe: Normal and psoriatic. *Bioengineering and Skin* 3:118 (1981).
42. Lykken, D. T. Properties of electrodes used in electrodermal measurement. *J. Comp. Physiol. Psychol.* 52:629–634 (1959).

8
Techniques for Sampling the Bacterial Flora of the Skin

ANTHONY A. GASPARI* *Emory University School of Medicine, Atlanta, Georgia*

INTRODUCTION

It is well known that the skin is permanently colonized by bacteria. Until recently, there have been few studies that address the issue of the distribution, location, types, and numbers of bacteria that inhabit the skin. The original work for microbiological sampling of surfaces began in the 1930s when scientific interest was directed toward the sampling of inanimate objects, such as multiple-use eating utensils, as a possible vehicle for the transmission of bacterial and viral infections. It was not until the late 1950s that it was realized that there was a lack of information concerning the microbiology of the skin, at which time investigators began to study the properties of several skin disinfectants that had been formulated to control nosocomial infections.

---

*Present affiliation: Strong Memorial Hospital, University of Rochester, Rochester, New York.

Since that time there has been a flurry of activity in the field, with the evolution of five specialized categories of sampling techniques to determine both quantitatively and qualitatively the microorganisms that reside on and in the skin. These categories include: (1) impression techniques, (2) washing and scrubbing techniques, (3) swabbing techniques, (4) biopsy techniques, and (5) air sampling for "shedders." (See Table 1 for a descriptive summary.)

In 1978 Staal and Noordzij (1) outlined the characteristics of what they considered to be an "ideal" sampling technique. First, the method should be easy to use. It should not be too cumbersome or time-consuming for field studies, and it should be amenable to processing large numbers of specimens. The results obtained should be reproducible, with minimal inherent error in the measurements. Investigators using this method should be able to obtain results comparable to others' results using the same method. Additionally, an individual should be able to apply this technique and obtain some internal consistency to his results.

This idealized method should be nontraumatic to the subjects being studied. This consideration becomes particularly important when dealing with human subjects in a situation that requires multiple, repeated specimens from one individual. Finally, this sampling technique should be applicable to all areas of the body. This factor may not be critical when sampling the flora of a flat skin surface that has only vellous hairs and a low density of microbial colonizers, but it acquires particular importance when samples must be taken from areas that are heavily populated with telogen hairs or from intertriginous areas that present folded skin surfaces with a high density of microbes. It is easy to understand that it is difficult to gain access to these areas for adequate sampling.

Investigators have applied creative modifications to the basic techniques in order to fulfill their experimental needs or to correct a perceived deficit in an existing technique. This had led to a proliferation in the number of available sampling techniques in the last 30 years, with the net result being that the sampling methods are becoming more specialized, all of which suggests that there is no one sampling method that is universally applicable.

There is a wide divergence of data concerning quantitative aspects of microbial populations of the skin. There are conflicting reports concerning microbial density at a given site, and there are also large sampling variations recorded by individual investigators using a single sampling technique. For example, Canuto (2) found only 253 organisms/cm^2 on the hands, while Price's figure (3) for the same area is 3200 bacteria/cm^2, and Arnold (4) reported a density of only 170 bacteria/cm^2.

Many investigators offered explanations for this phenomenon. Price (3) felt that the variability is secondary to the diversity of the skin flora, rather than technical error. Pachtmann et al. (5) suggested that day-to-day variations of the individual flora are responsible. Noble (6) studied the cutaneous populations of 22 "normal" individuals and found that the log of the numerical counts for a certain sampling site is distributed in a normal fashion. There are individuals with persistently high and persistently low counts, and these individuals' counts may differ from those of the rest of the population in a statistically significant manner.

Evans et al. (7) pointed out that there are marked differences in the bacterial populations within a few centimeters. Their bacteriological study of normal skin also demonstrated that counts vary tremendously from individual to individual and that counts taken on a single individual from a single site will widely fluctuate (when followed serially). Ulrich (8), in 1964, reviewed the existing sampling techniques and implied that the inherent faults of each method were responsible for many of the reported discrepancies.

Let us now examine the basic techniques in detail: how they were developed and modified; their strengths and weaknesses; and their applicability in clinical and investigative situations.

IMPRESSION TECHNIQUES

The idea that led to the development of impression culturing methods was originated by Lederberg and Lederberg (9) in 1952, when they used sterilized velvet or velveteen pads to transfer bacteria from one agar plate to another. They obtained perfect replication of colony relations between plates. Their method was originally intended to isolate bacterial mutants and to study biochemical reactions and antibiotic sensitivity.

Holt (10), in 1966, used sterilized velvet velour pads (with aluminum backing to facilitate their handling) to sample the skin. He wet his pads with sterile nutrient broth and then applied them to the skin and, after a few seconds, carefully peeled the pads off. They were then overlaid onto the surface of a suitable nutrient agar. The advantages of his method include the ability to make replicate plates with the velvet pad and then use selective media. Holt suggested that his method be used in a clinical setting for studying the bacteria around surgical wounds, for examining the survival of experimental pathogens on the skin, for studying colony maps of the skin and the flora of hard surfaces, or for the investigation of sterilization of the skin.

TABLE 1 Summary and Description of All Major Sampling Techniques

Technique	Ease of use	Reproducibility	Traumatic	Microcolony size
Impression methods (6,9,10−14, 19−23)	Good "field" method; rapid and simple	Good (12,14)	No	Aggregated unless using Raahave's modific (12)
Washing methods (3,25−27)	Cumbersome; requires serial basins (3)	Fair	No	Dispersed
Scrubbing methods (5,28−30, 32−35)	Involves special equipment	Good; a widely used method	No	Dispersed
Swabbing methods (6,8,14, 38−45)	Fast, simple; no special equipment	Poor, not a quantitative technique (6)	No	Dispersed
Biopsy methods (16,37,47, 48,50−52, 54−69,63, 64)	Always involves special equipment	Good	Yes, but can be minimized with modific	Superficial; subsurface; deep
Dispersal techniques (6,71−75)	Requires specialized equipment	?	No	Intact on disquamated corneocytes

Flora access	Advantages	Disadvantages	Miscellaneous
Surface flora only; subsurface if using modific (12)	Good field method; easy to make replicate plates	Cannot use in areas where counts are high; superficial access; inefficient sampler	Counts subject to technical error; can modify tech. to study lab model of contact transfer (23)
Surface and subsurface	Useful as a lab model for "degerming" hands (3)	Wash solutions probably bacteriostatic and decreases counts	Especially useful for studying bacteriology of hands; disperses corneocytes
Surface and subsurface	Most convenient and reproducible method	Colony counts affected by variables such as force, rotation of spatula; difficult to use in intertriginous areas	Dispersal fluid can affect colony counts (30, 31); mechanized (33) and ultrasound (34) variations enhance yield
Surface and subsurface	Good field method; useful for qualitative microbiology; useful for processing large numbers of specimens	Yield influenced by swabbing time, pressure, moisture content of swab head	Some swab heads bactericidal, bacteriostatic; wetting agent affects counts; inefficient sampler
Superficial; subsurface; deep	Can visualize in situ distribution of microcolonies; access to all levels of flora	Inflicts trauma, adhesive of skin stripping methods is bacteriostatic (49–52, 58–60)	Particularly useful in the study of hair follicle flora (relevant to acne studies) (54–56)
Superficial only	Useful for studying large volumes of air in a short period of time	Airborne contaminants are unsettled question	Great relevance in the study of transmission of infections

In 1962, Green et al. (11) applied floc material* to a section of metal pipe, and after sterilization it was applied to nutrient agar and then rolled one or more times over the surface to be sampled. It was then rolled onto fresh, sterile nutrient agar for culturing. This method was used to study microbial colonization of inanimate objects, but it has potential for application to the study of cutaneous microflora.

Raahave (12), in 1975, studied the efficiency and precision of bacterial recovery from the skin using the velvet pad impression method. In his study he examined the recovery of a known number of organisms inoculated onto sterile agar. He found that the uptake of the surface microbes was close to 100%, but the release of the organisms onto fresh agar wash was only 1–4%. By rinsing the pads with sterile saline after contact with the test surface, then centrifuging the eluent, and then inoculating the sediment on sterile agar, he was able to increase his yield to about 60% of the original inoculum.

This treatment, which improves his yield, results in a disaggregation of viable colonies into individual microbes. When unmodified impression techniques are used, the bacteria that are sampled are in the form of undisturbed viable aggregates. By dispersing these aggregates, the resulting bacterial counts are increased because colonies are disrupted and transformed to individual cells. For this reason, dispersal techniques are not comparable to impression techniques that do not disrupt microcolony aggregation.

The problem of loss of organisms by transferring from the sampling device to the incubating medium can be eliminated by using a direct sampling technique. (Up to this point, we have been discussing indirect impression methods.) This direct method involves using the culture (and ultimately the counting) medium as the sampling device. The agar can be poured into a container (usually a Petri dish) so that the meniscus rises above the edge of the dish, and this is applied directly against the skin or surface to be sampled. The medium, along with the container, can be varied, depending on the surface to be sampled and the organisms the investigator wants to isolate. In 1965, ten Cate (13) used polyamide casings filled with agar to make "agar sausages"

---

*Floc material is pulverized cotton, wool, silk, or rayon fibers that form a velvety pattern on cloth.

that were then pressed to the surface to be sampled; then sections were sliced away and placed in a Petri dish for culturing. A new, sterile surface of agar will then be exposed for further sampling. Favero et al. (14), in their review of the available techniques for sampling surfaces, described some other direct impression techniques. One such method is the direct agar plating technique, wherein liquefied agar can be poured directly onto the surface to be sampled and allowed to solidify. The slab is covered by a sterile cover and later removed for sampling. They also elaborated on using a syringe or a syringelike tube filled with media for direct sampling of the skin.

Impression methods are a good "field method" because they are simple to use and are thought to yield reproducible results (6,12,14). This technique samples intact microcolonies or aggregates of viable surface bacteria, and this has numerous implications. First, it is of limited usefulness where the density of organisms is high because no dilutions are done, and the confluence of colonies can make counting impossible. The presence of rapid growers or "spreaders" will also adversely affect the counts. Imprecise sampling or inoculation will result in smearing of colonies, again adversely affecting the colony counts. Any disruption of microcolonies by sampling or inoculation that results in the release of individual cells will lead to erroneously high counts since these individual cells will grow and form colonies. Additionally, the usefulness of impression techniques is limited to smooth, flat surfaces.

Mary Marples (15), in 1965, suggested that the cutaneous bacteria resided in the pilosebaceous unit but were also found in lesser numbers scattered over the surface of the skin. Montes and Wilborn (16) confirmed this idea in 1969 by actually visualizing the distribution of bacteria using scanning electron microscopy (SEM). They noted more numerous organisms at the openings of hair follicles than on the general skin surface. In a refinement of their first study, Montes and Wilborn (17) also demonstrated a dense, rather heterogeneous population within the infundibulum of the hair follicle. This reservoir of organisms, which actually constitutes the majority of the total skin flora, is inaccessible to the surface impression techniques since they sample only superficial colonies.

Additionally, Marks and Dawber (18), in 1971, used SEM to visualize the uneven topography of the microenvironment of the skin. The numerous "hills and valleys" of clinically smooth, glabrous skin provide numerous recesses that cannot be adequately sampled by a surface method. Impression techniques that involve wetting may minimize this problem but, at the same time, may disrupt intact colonies.

Other impression techniques that have been used in cutaneous microbiological studies include a "finger streak" test (19) to evaluate disinfection methods of the hands. As the name implies, the fingers of the hand are gently drawn across the surface of sterile agar plates. There is a great deal of variance of results, and frequently the colony counts are impossible to do because of the confluence of colonies.

Hair can be sampled by impression techniques. Noble (20), while studying *Staphylococcus aureus* incidence in the hair of normal subjects and patients with skin disease, obtained his cultures using both direct and indirect methods: that is, a velvet pad technique and a direct contact agar plate. The direct contact method yielded consistently higher counts than did the velvet pad technique. Summers et al. (21), in 1965, conducted a similar study and obtained higher counts with a direct contact technique.

The hair brush, although used primarily for diagnosing dermatophyte infections of the scalp, can also be applied to bacteriological sampling of the hair (22). It is an inefficient method when compared with other impression techniques.

Impression methods have been ingeniously applied to the study of a laboratory model of contact transfer of bacterial organisms by Marples and Towers (1979) (23). Their study was designed to assess simple means of degerming the hands. They covered a bottle with fabric that was contaminated with known numbers of *Staphylococcus saphyrophyticus*. The subjects would grasp this bottle, then wash their hands, and afterward would grasp another bottle covered with sterile cloth. The fabrics were then removed from the bottles and agitated in a jar with 100 ml of Triton-X 100 and were then serially diluted for culturing. The study demonstrated that the key factor in transferring organisms from bottle to hand was the presence of moisture. Different materials also had marked effects on the transfer of organisms. Minor procedures as simple as wiping the hands significantly decreased the number of organisms transferred.

WASHING TECHNIQUES

Washing methods disperse aggregated bacterial microcolonies to individual cells and also disrupt corneocytes. McBride et al. (24) counted epithelial cells present in the wash fluid using a cup scrub method and attempted to correlate this with the number of organisms isolated. His group found that the correlation coefficient for the number of epithelial cells and the total number of bacteria was 0.65 but was much lower for individual species (except for the propionibacteria), which had a correlation coefficient of 0.64). This suggests that large populations of anaerobes are associated with epithelial cells and that the physical properties of the stratum corneum may contribute to variations in bacterial populations. Thus, washing techniques appear to have better access to the deeper microbial flora, especially the anaerobes.

Price (3), in 1938, developed the serial basin technique to evaluate degerming of the hands. With this method, the hands are scrubbed uniformly in a series of basins, and aliquots are taken from each basin and inoculated onto agar medium for incubation and later colony counting. He plotted the log of the counts from the basins against time and obtained a smooth "degerming curve" for the hands. His original technique and its modifications (25,26) are still widely used, although they are cumbersome and are applicable only to the hands. These techniques are useful for laboratory and classroom instructions, as they nicely demonstrate the slow, steady, yet incomplete degerming of the hands with serial washes.

Another group of investigators (27) used a sterile glove to sample the flora of the hand. After the hands are washed with the degerming agent to be studied, a sterile glove is applied to the hand, and sampling fluid is placed inside the glove. Exposure of the sampling fluid to the skin surface is maximized by squeezing the gloved fingers for a predetermined period of time. When this is finished, an aliquot of fluid is removed and added to a neutralizing solution (lecithin-phosphate buffer) to destroy any residual antibacterial solution left on the hands after washing. The neutralizing solution is the key, for any bacteriostatic or bactericidal chemicals that find their way onto the culture medium will falsely lower the colony counts. This was surely a problem with Price's original technique, as he used soapy water to inoculate onto the culture medium.

In 1938, Burtenshaw originated the scrub technique (28). His sampling chamber consisted of a rectangular jar with the bottom cut off, his sampling fluid was a saline solution, and he used a glass slide as a "scrubber." He originally developed this technique to study the survival of *Staphylococcus* species on skin and inanimate objects. Despite many refinements and modifications, his basic technique is still recognizable today.

Pachtmann et al. (5), in 1954, modified Burtenshaw's scrub method for their study of cutaneous flora. They used a sterile glass cylinder pressed against the skin with a 2-ml volume of wash fluid (BHI) placed within it. The skin enclosed by the cylinder was then scraped 20 times with a sterile wooden spatula. The sampling fluid was then serially diluted and plated for subsequent counting. Although this technique produces more reproducible results than the serial basin technique (and its variations), Pachtmann et al. still encountered large inter- and intra-individual variability in their colony counts.

Williamson and Kligman (29) refined Pachtmann's technique by substituting a sterile teflon "policeman" for the wooden spatula. They applied two separate sampling aliquots as washing fluid and eventually pooled the two washes together for serial dilution and plating for subsequent counting.

Their most important modification of this technique was the demonstration that the wash fluid had a significant role in dispersing colonies to individual cells, thus increasing their yield. They chose Triton-X 100 (0.1%) in phosphate-buffered saline (pH 7.9) because it was thought to be nonbactericidal, had low irritancy properties, and would be a good dispersant. After doing bacterial survival studies in Triton-X 100, Bloom (30) suggested plating the samples within 30 min after obtaining them because Triton-X 100 has no nutrients, and this detergent solution may impose the added stress of disrupting cell membranes and thus may adversely affect the growth of both aerobes and anaerobes. Kishishita et al. (31) subsequently demonstrated that Triton-X 100 may be bactericidal for anaerobes (although it is suitable for aerobes) and suggested using 0.1% Tween-80 in phosphate-buffered solutions when investigating the propionibacteria.

Marples et al. (32) further tailored the Kligman technique by applying the convenient "drop inoculation" procedure instead of his incorporation plates. Drop inoculation involves sampling the skin with swabs moistened with 0.1% Triton-X 100 and phosphate-buffered saline (pH 7.9), returning the swab to 2 ml of dispersal fluid, and then mechanically agitating the swab; serial dilutions are further done, and finally, a drop of the dispersal fluid is inoculated onto solid medium for culturing.

Bibel and Lovell (33) used a reciprocating instrument to add further precision to the wash technique. This mechanized device standardized the number and force of the strokes of the spatula used to sample an area within a stainless steel square section that had been applied to the skin.

Stringer and Marples (34) applied ultrasonic impulses to the detergent wash technique to speed up the disaggregation and removal of bacteria. Staal and Noordzij (1) used a pulsating water-jet spray device (Water Pik) attached to a sampling chamber, while Thran (35) used a continuous-spray device to obtain bacteriological specimens.

Hartmann (36) compared Williamson and Kligman's buffer scrub method to the spray techniques of Staal and Noordzij (1) and Thran (35) and found that the buffer scrub consistently elaborated more colony-forming units and more aerobes and anaerobes than the spray methods. The two spray methods were equivalent in efficiency; all of these wash methods were more efficient than a skin stripping method (37).

Bibel and Lovell (33) effectively used a buffer scrub technique in their studies of cutaneous ecology using skin flora maps. These "maps" consist of 10×10 cm^2 grids drawn over the entire surface of the skin; a sample is taken from the center of each grid. This imaginative technique is useful for studying the total skin flora, but only one or two subjects can be studied at a time because of the tremendous numbers of samples that are generated by using this technique.

SWABBING TECHNIQUES

Obtaining skin cultures using the swabbing technique is a common practice, mainly because the method is fast, simple, and amenable to field use where there will be large numbers of samples to obtain and process.

It is well known that swabbing techniques are semiquantitative; some authors consider them qualitative at best (6). There is at times a poor correlation between the number of organisms present on the surface and that recovered by investigators. Noble estimated that only one one-hundredth of the total number of organisms on the skin surface are transferred to the agar culture surface for subsequent counting. Results reported by different investigators can become uncomparable if those doing the study are not cognizant that bacterial recovery can be affected by swabbing time, swabbing pressure, and the moisture content of the swab (8).

Additionally, Hucker et al. have shown that the yield of bacteria obtained by swabbing is directly proportional to the ease of wetting the surface to be sampled (38). The material that composes the swab head and the wetting solution affect the bacterial recovery rates by their potential bacteriostatic or bactericidal effects.

Anderson (39) noted that bacterial survival on swabs was affected by pH and suggested using a wetting agent with a pH of 7.0 to maximize bacterial survival. Douglas (40) confirmed the increased bacterial recovery by moist swabs and further noted that the choice of the moistening agent enhanced or depressed colonies isolated. He noted that Ringer's solution was toxic, while peptone broth was not. Shaw et al. (41) wet swabs with 0.1% Triton-X 100 and, using a rayon swab, were able to obtain quantitative results similar to those of a cup scrub technique. They also noted the smallest standard error when using this swab and wetting technique.

Rubbo and Benjamin (42) studied survival rates on cotton swabs moistened by various methods and compared a serum-pretreated cotton swab (swabs were rolled and dipped in ox serum and dried at 37°C for 1 hr) with an untreated cotton swab. They found that the serum-treated swabs yielded heavier growth and allowed prolonged survival compared with the moistened, untreated swabs. Moisture increased the death rate of gram-positive organisms on the plain swabs but not on the serum-treated swabs. Other colloids (such as albumin and ether extracts of serum) that were substituted for ox serum also had protective effects on bacterial survival.

Another study (43) investigated the value of using desiccated swabs for streptococcal epidemiology in field studies. Throat and skin cultures were obtained in duplicate with calcium alginate swabs. One was plated immediately onto selective media, and the other was placed in an aluminum foil packet with sterile silica gel and was plated in 4 weeks. The inestigators found that there was good correlation between the colony counts obtained from swabs plated immediately and those plated in 4 weeks. This correlation between immediate and delayed plating held true for the skin swabbings, but not for the throat swabbings.

The effect of the composition of the swab on bacterial survival and recovery has also been studied intensively. Cotton wool swabs are widely used, but Higgins (44) suggested calcium alginate swabs because she obtained consistently higher colony counts in her comparison study of the two types of swabs.

Calcium alginate swabs dissolve in sodium hexametaphosphate, or Ringer's solution, thus freeing the sampled bacteria into the plating solution. Favero et al. (14) suggest that the calcium alginate swab does not remove as many organisms from the skin as the cotton wool swab, but other investigators (45) feel that the calcium alginate swabs are as effective as the cotton wool swabs in removing bacteria from the skin.

Shaw et al. (41), in 1970, compared rayon with calcium alginate swabs and found that while their bactericidal properties were similar, rayon swabs were more effective than calcium alginate in recovering organisms.

All of these studies are suggestive and at times contradictory and inconclusive, and it remains to be clarified which swab is most efficacious in sampling organisms from the skin surface, which is the least bactericidal, and which maximizes the release of these organisms onto the surface of the culture medium for subsequent counting.

SURFACE TECHNIQUES

Up to this point we have been considering only "surface techniques" for sampling the skin. The scrub, wash, and impression methods do not truly sample the deeper flora of the sebaceous follicle, only those organisms which "spill" out of the hair follicle onto the surface of the skin. The deeper follicular flora are not available using these surface techniques.

Noble (6) has compared the efficiency of swabbing with scrubbing and contact culture techniques. He used all three methods in parallel on six different subjects in three different areas of the body and obtained consistently high counts with the scrub technique, followed by the swabbing method, and last, the contact method.

These results are not surprising, since both the swab and scrub methods tend to disrupt the stratum corneum and loosen corneocytes (24) with their adherent flora. Also, the scrubbing and swabbing techniques are dispersal techniques that elaborate individual bacteria from an aggregated colony, whereas the contact method samples intact colonies without dispersing individual cells.

By combining surface techniques, the information gained from one method may complement what is lacking in another technique and thus give a better perspective to understanding the skin flora. Evans and Stevens (46) did such a study in 1976. They compared the colony counts obtained from the palms (no sebaceous glands and therefore only surface flora) using both a swabbing and a scrubbing technique. Both techniques were equally effective in sampling the bacterial flora in this area. On the forehead, where the density of sebaceous glands is 400–900 cm^2 (15), the scrub technique consistently obtained counts five to eight times higher than the swab technique. (This is only when counting anaerobic *Propionibacterium*; the counts of surface organisms on the forehead were much too erratic to make any comparisons about efficiency.)

One distressing problem that arose in this study was the variability of the number of organisms obtained when sampling adjacent sites. The authors felt that this effect "reflects conditions on the skin where bacteria grow in microcolonies and are distributed unevenly, not as a diffuse, evenly mixed flora."

This study clearly demonstrates that swabbing is an efficient method for sampling surface organisms, whereas scrubbing is effective for sampling surface and subsurface organisms. When these two methods are used in series (swabbing and then scrubbing methods) on the forehead, the scrubbing and swabbing techniques fill in gaps of the other technique. The swabbing method delineates almost exclusively surface organisms and leaves undisturbed the subsurface flora, whereas the scrubbing method will effectively sample both sources of flora. Neither method will effectively obtain organisms from the deep follicular flora.

BIOPSY TECHNIQUES

Biopsy techniques offer the advantage of sampling *all* of the bacteria present in a given area of skin. These methods are thought to be an effective way of examining qualitatively and quantitatively the in situ microflora of the skin. The biopsied specimen can be stained or cultured after it is obtained, and the sampling artifacts that are frequently problems with other sampling methods are not a difficulty when dealing with this technique.

It is frequently mentioned that biopsy methods inflict trauma to the experimental subject, and this can be a problem with excisional and punch techniques, but as we will see in the following passages, obtaining a biopsy for bacteriological processing using new techniques does not necessarily entail pain or disfigurement.

Some of the most revealing data concerning the localization, quantitation, and distribution of the skin flora have come from bacteriological study of skin biopsies.

Lovell (47), in 1945, did pioneering work by culturing skin samples obtained surgically and then fixing them for staining procedures. He was the first to demonstrate the hair follicle as a place of bacterial residence and failed to find any organisms in the sweat gland. He concluded from his study that surface bacteria can be cleaned off, but resident bacteria in the hair follicle cannot be removed without damaging the skin.

In 1969, Montes and Wilborn (16) used skin biopsies for both light microscopic (using PAS or H&E stains) and electron microscopic studies to directly visualize the anatomical location of bacteria in the pilosebaceous unit and its absence in the sweat gland, thus confirming Lovell's work in 1945.

Baxby and Woodruff (48) used a Castroviejo keratotome (a mechanized ophthalmological surgical instrument) to obtain uniformly thin layers of skin from pigs and palmar human skin. The samples were disintegrated and then weighed, serially diluted, and cultured. These authors were able to make multiple skin slices of constant thickness (correlated by weight) from a single area. They found that with pigskin their highest counts were in the surface 0.3 mm of epidermis. (Because of the smaller "slices" of the human skin that were taken and the variability of their results, they could not comment on bacterial density in the human.) They concluded that, at least in the pigskin animal model, bacteria are situated in the stratum corneum and speculated that this was related to the presence of sebaceous and sweat glands.

Other investigators have centered their attention on the microflora of the pilosebaceous unit. Holland et al. (37) used cyanoacrylate gel (Permabond) to sample pilosebaceous units. After the gel is placed on the skin (within the confines of a sterile teflon ring), a glass sampler is pressed to the skin and then removed with the adherent skin biopsy. The sample is then spun in a blender with glass beads and dispersion medium (to disaggregate organisms) and then incubated. Holland's group claims this method to be a subsurface method, as they recovered less than 1% of a surface marker organism. The authors feel that this method's advantages are that it is painless, that the number of pilosebaceous units per sample can be counted under low power, and that their method is useful for quantitative and qualitative studies. This method is a dispersal technique that measures individual cells as colony-forming units.

Permabond has bacteriostatic and bactericidal activity against *Corynebacterium acnes* and *Staphylococcus albus*. This limits the applicability of this adhesive since it is well known that both *Corynebacterium* and *Staphylococcus* are ubiquitous, present in large numbers at most skin sites that bear pilosebaceous units in normal individuals (49).

Kooyman and Simons (50) used sterilized cocktail picks with a drop of adhesive on them to biopsy the skin for microbiological studies. After a 36-hr incubation period on agar medium, they counted colonies under the disk area only. Although their method is simple and rapid to apply, it cannot be used on moist, weeping skin (as with other adhesive techniques), and there is, again, the problem of bacteriostasis with the adhesive material. Since the skin sample is placed between the agar and the sampling device for incubation, aerobic colonies grow slowly, are small, and are difficult to count. Additionally, this method samples aggregated colonies, so it is not really comparable to Holland's adhesive biopsy technique, which is a dispersal method.

Malcolm and Hughes (51) biopsied the skin of the sole of the foot (with and without previous occlusion) in preparation for scanning electron microscopy. They obtained the biopsies by using a needle and razor blade or with the cyanoacrylate adhesive. Both biopsy methods were equally effective in showing the few scattered bacteria on the surface of the unoccluded skin and the increased numbers of bacteria congregrated around the opening of the sweat ducts of the occluded skin.

Perhaps the adhesive method is most useful when performing static, morphological bacterial studies, rather than quantitative or qualitative studies that involve culturing bacteria. Marks and Dawber (52) dropped their adhesive directly onto the skin and pressed glass slides to obtain the specimens, which were then ready for special stains and permanent mount. They also used this technique to study invaders of the horn (53) and made permanent mounts of ringworm, erythrasma, pitted keratolysis, and impetigo.

Whiteside and Voss (54), in their biochemical study of the lipolytic activity of *Corynebacterium* in normal and acne skin, expressed the contents of pilosebaceous units with a comedone extractor. They then streaked the contents on agar for culturing (with and without prior homogenization). They found that normal sebaceous follicles of the nose are essentially pure cultures of *Propionibacterium*, with *Staphylococcus* species being found infrequently in the normal follicle. Another acne study (55) utilized the same technique but separately pooled five open and closed comedones for qualitative microbiological studies.

Although useful in acne studies, this method of examining the flora of the pilosebaceous unit is limited to areas where there are enlarged follicular openings or comedones. It is not useful in other types of studies.

Sampling Bacterial Flora of Skin

Puhvel et al. (56) used biopsy techniques to isolate individual pilosebaceous follicles. After a routine 3.5-mm punch biopsy of the skin, the specimen was incubated in 1 molar $CaCl_2$ at 4°C for 1 hr. The epidermis can then be separated from the dermis and the sebaceous glands cut off and ground up for subsequent microbiological processing. Puhvel et al. found that the $CaCl_2$ solution was not bactericidal, but there was a loss of organisms into the dissecting fluid (averaging about 10% of the total yield). After weighing the sebaceous follicles and performing quantitative microbiological studies, they found the correlation coefficient between follicle weight and bacterial numbers to be 0.6.

Mustakallio et al. (57) used a biopsy technique to study surface and subsurface flora. They obtained epidermal specimens by inducing suction blisters (by 3 hrs of vacuum 150–200 mm Hg below atmospheric pressure), which results in a clean separation at the dermal-epidermal junction. The epidermal sheet is weighed, homogenized in buffered Triton-X 100, and then cultured. These studies suggest that neither the homogenization nor the buffer solution affect the counts, but these authors failed to consider that 3 hrs of barotrauma may have adverse effects on the survival of aerobic organisms in the epidermis.

Mustakallio et al. point out in their study that other dispersal techniques (such as scrubbing and scraping) rely on the dispersal medium at the time of sampling. Penetration of the sampling fluid may be an uncontrollable variable that introduces error in sampling. This problem is circumvented in Mustakallio's technique by exposing the sample to dispersal medium after the sample is obtained.

The suction blister technique is useful because it enables one to obtain a full-thickness sheet of epidermis without pain or scarring. As with other biopsy techniques, all of the bacteria are removed with the sample, so bacterial counts can be related to the biopsy area and/or weight. By itself, the suction blister technique is limited because of the superficial access that it attains. This method, when used in conjunction with Puhvel's method (56) for isolating the pilosebaceous unit, would be useful in comparing the superficial and deep inhabitants of the pilosebaceous unit.

Adhesive tape is a simple, useful tool for studying the biology and pathology of human skin. Wolf (58,59) used Scotch brand cellotape to study the structure of keratinocytes and found that successive layers of the horn could be removed by adhering the tape to the skin and then quickly peeling it off. Pinkus (60) used Scotch brand 600 tape to remove 20–30 layers of cells to study epidermal mitotic rates.

Tape strippings can be used to diagnose dermatophyte infections of the skin and are thought to be as effective as scraping the skin for obtaining cultures, although not as effective for KOH preparations for microscopy (61). Tape stripping is facilitated by hydrating the skin before the samples are taken. Weigand and Gaylor (62) demonstrated that hydrating the skin results in less erythema of the sripped area, as well as less horny layer remaining.

Tape stripping has also been used to study the bacteriology of the skin. Although this method is rightfully classified as a biopsy technique because it necessitates the removal of host tissue to sample its flora, in many ways it is similar to the contact method. Updegraff (63) compared tape stripping with the wash method of Price (3) and the scrub method of Evans (7) and Pachtmann (5) and found that his counts were consistently lower than those of the other methods, suggesting that the tape strip method elaborates intact colonies rather than individual organisms.

The intact microcolonies, which are elaborated along with the horny layer, must then be deposited onto a culture medium. None of the studies reported have attempted to disperse these microcolonies before incubation, so the tape is either covered with agar, or it is placed onto the surface of the agar for incubation. In either case, the resulting colonies will be placed in a milieu with decreased oxygen tension (since the organisms are "sandwiched" between agar and tape), so aerobic colonies will grow poorly, if at all.

Another difficulty with the tape strip method is the bacteriostasis that occurs secondary to the adhesive film and plastic backing. Updegraff (63) studied over 25 different tapes for their bacteriostatic activities and found that all tapes were bacteriostatic, some more than others. He has developed a culture medium that minimizes this problem when using certain tapes. Some tapes evan elaborate a volatile component that is bacteriostatic, even when taped to the outside of the Petri dish (64).

The usefulness of the tape strip method lies in its simplicity and its ability to sample in a serial fashion the depth at which organisms are found in the epidermis. As with the acrylic adhesive methods, it is most useful for fixed permanent specimens rather than for cultural studies.

SPECIAL SITUATIONS

When studying the bacteriology of lesional skin (i.e., burns, infections, wounds, psoriasis, etc.), special care must be taken not to induce pain, retard wound healing, exacerbate or induce an infection, or in any way worsen the subject's preexisting morbid state. Washing (65), impression (66), swabbing (67,68), and biopsy techniques (69,70) have all been used to sample bacteria from abnormal skin after modifications have been incorporated to meet the demands of a delicate situation.

AIR SAMPLING

When sampling the dispersal of bacteria into the air on desquamated skin scales, the subject is usually enclosed in a cubicle and is asked to undress or to perform exercises, which is known to disperse large numbers of scales and bacteria (71).

The skin particles that are shed (with their adherent flora) can be sampled by using settle plates, which are simply agar plates that are left exposed to the air for set periods of time. This method is simple and inexpensive and relies on gravity to bring the desquamated samples into contact with the sampling medium.

While settle plates are a simple and inexpensive way of sampling the air, they are inefficient because long exposure times and a large sampling surface area are required for adequate volumes of air to contact the plates (6).

Impaction slit samples are an efficient, mechanized way of sampling large volumes of air in a short period of time. This device consists of a slit that is exposed to the air of the cubicle in which the subject is performing activities that will disperse scales and bacteria. Below the sampling slit is a chamber that contains a sampling plate that holds either solidified agar or glass slides coated with petrolatum jelly. It is upon these surfaces that the particle-laden air comes to deposit its contents. The plates are then incubated in a routine manner, and qualitative and quantitative microbiological studies are performed. The volume of air drawn into the chamber is controlled by a fan, while the surface area exposed for sampling is regulated by a mechanized turntable, upon which rests the sampling plate.

This device is widely used in the study of bacterial dispersal (72--74), particularly *S. aureus* (75).

RECOMMENDATIONS

When deciding which sampling techniques to apply to an investigative situation, the objectives of the study must be well planned.

For example, in testing the effects of an over-the-counter product on the survival of propionibacteria in subjects with acne, several factors must be considered before choosing a sampling technique. First, the geographical area under study is to be considered. Since the areas of skin involved in this disease process are limited to the face, chest, or back, these areas are commonly examined.

Biopsy techniques are useful in the quantitative study of propionibacteria because they will adequately sample both the superficial and deep flora, and the data are reproducible. If the area of interest centers on the face, then it is not advisable to obtain bacteriological specimens using the punch or excisional biopsy techniques.

In the study of the flora of the back or chest where biopsy scars are more cosmetically acceptable, such aggressive biopsy techniques are a reasonable approach to the study of propionibacteria, since the biopsies can be processed using Puhvel's technique (56) for isolating individual pilosebaceous units.

It is possible to obtain a superficial biopsy using the adhesive glue methods (37,50--53), but the sample is not amenable to quantitative cultural studies but more useful for obtaining permanent fixed sections. Furthermore, I have previously alluded to the exquisite sensitivity of the propionibacteria to the cyanoacrylate adhesives.

A technique that is especially useful in the bacteriological study of acne, especially the propionibacteria, is that of Whiteside and Voss (54), who expressed the contents of the pilosebaceous unit using a comedone extractor. This is easily applied to the face, where there are large follicular openings or areas where comedonal lesions exist.

Impression methods are not useful in the study of acne because of their limited superficial sampling access; these methods cannot sample the deep follicular flora, the predominent place of residence of the propionibacteria.

Swabbing methods are popular because of their simplicity and may be useful for qualitative studies of heavily colonized or infected tissue. This method has limited usefulness for quantitative work because it is an inefficient sampler, and some of the sampling heads may be bacteriostatic for the sampled microbes. Williamson and Kligman's technique (29) is widely used because of its relative simplicity, reproducibility, and access to the subsurface flora. I do not unhesitatingly recommend their method; it is absolutely necessary to use a dispersing agent other than Triton-X 100, possibly Tween-80. It is advisable to perform preliminary in vitro bactericidal/bacteriostatic studies using a variety of wetting agents to minimize this problem.

When examining the microbiological flora of special areas such as the axilla, the toe webs, or other intertriginous areas that are heavily populated by microbes and from which it is difficult to obtain samples, the usually efficient cup scrub is not useful. The reason for this incompatibility is related to the lack of a flat sampling surface, which is necessary in order to form a tight seal with the sampling cup. In this situation, the pulsating water spray method of Staal and Noordzij (1) and the continuous water spray method of Thran (35) are particularly useful. Their small, efficient sampling chambers are well suited to these difficult areas, although the buffer scrub method consistently elaborates higher colony counts when sampling flatter skin surfaces (36).

For routine surveillance of the flora of glabrous skin as related to over-the-counter product use, impression methods are adequate because of their simplicity and reproducibility, but it is important to note their superficial access. For the study of subsurface flora, the buffer scrub is most effective for nonspecialized flat skin surfaces.

The hands are an exception, in that buffer scrub techniques offer no advantages over simple impression techniques. Palmar and plantar skin is devoid of sebaceous glands, so there is no deep follicular flora. The flora of the hands is superficial and frequently consists of saprophytes, gram-negative rods, and other transient inhabitants of the skin. Washbasin techniques such as the modified Price technique (25–27) yield useful and reproducible data but are rarely utilized because they are cumbersome and time-consuming. Thus, impression methods may be useful in the study of palmar and plantar flora, especially in the routine daily surveillance of the effects of use of over-the-counter products.

This discussion has described the development, uses, and pitfalls of the major techniques available to sample the skin. It is by no means all-encompassing, as the proliferation and variety of methods available make this a cumbersome task. The very existence of such a melange of techniques suggests that there is no one "ideal" method.

A comment made by Ulrich in 1964 (8) places this subject in its proper perspective: "Each sampling method determines a different aspect of microbial skin populations, and they supplement one another in the data derived from them. No one method gives a rounded view of the dynamics of the microbiology of the skin."

Selwyn and Ellis (76) add, "It is a fallacy to believe the total resident flora consists of those organisms accessible to sampling methods."

We must consider these admonitions when we are deciding which sampling method to use for investigational or clinical activities or when interpreting results presented by other investigators.

REFERENCES

1. Staal, E. M., and Noordzij, A. C. A new method for the quantitative determination of microorganisms on human skin. *J. Soc. Cosmet. Chem.* 29:607 (1978).

2. Canuto, A. Cited in Evans, C. A., Smith, W. M., Johnson, E. A., and Gilbert, E. R. Bacterial flora of the normal skin. *J. Invest. Dermatol.* 15:305 (1950).

3. Price, P. B. Bacteriology of normal skin. A new quantitative test applied to the study of bacterial flora and disinfectant action of mechanical cleansing. *J. Infect. Dis.* 63:301 (1938).

4. Arnold, L. Relationship between certain physicochemical changes in the cornified layer and the endogenous bacterial flora of the skin. *J. Invest. Dermatol.* 5:207 (1942).

5. Pachtmann, E. A., Vicher, E. E., and Brunner, M. J. The bacteriologic flora in seborrheic dermatitis. *J. Invest. Dermatol.* 22:389 (1954).

6. Noble, W. C. *Microbiology of Human Skin*, 2nd ed. Lloyd-Luke Medical Books, London, pp. 401–414 (1981).

7. Evans, C. A., Smith, W. M., Johnston, E. A., and Giblett, E. R. Bacterial flora of the normal human skin. *J. Invest. Dermatol.* 15:305 (1950).

8. Ulrich, J. A. Techniques of skin sampling for microbial contaminants. *Health Lab. Sci.* 1:133 (1964).

9. Lederberg, J., and Lederberg, E. M. Replica plating and indirect selection of bacterial mutants. *J. Bacteriol.* 63:399 (1952).

10. Holt, R. J. Pad cultues on skin surfaces. *J. Appl. Bacteriol.* 29:625 (1966).

11. Green, V. W., Vesley, D., and Keenan, K. M. New method for microbiological sampling of surfaces. *J. Bacteriol.* 84:188 (1962).

12. Raahave, D. Experimental evaluation of the velvet pad rinse technique as a microbiologic sampling method. *Acta Pathol. Microbiol Scand.* B 83:416 (1975).

13. ten Cate, L. A note on a simple and rapid method of bacteriological sampling by means of agar sausages. *J. Appl. Bacteriol.* 28:221 (1965).

14. Favero, M. S., McDade, J. J., Robertson, J. A., Hoffman, R. K., and Edwards, R. W. Microbiologic sampling of surfaces. *J. Appl. Bacteriol.* 31:336 (1968).

15. Marples, M. *The Ecology of the Human Skin.* Charles C Thomas, Springfield, IL, p. 202 (1965).

16. Montes, L. F., and Wilborn, W. H. Location of bacterial skin flora. *Br. J. Dermatol.* 81(Suppl. 1):23 (1969).

17. Montes, L. F., and Wilborn, W. H. Anatomical location of normal skin flora. *Arch. Dermatol.* 101:145 (1970).

18. Marks, R., and Dawber, R. P. R. Skin surface biopsy. An improved technique for examination of the horny layer. *Br. J. Dermatol.* 84:117 (1971).

19. Ayliffe, G. A. H., Babb, J. R., Bridges, K., Lilly, H. A., Lowbury, E. J. L., Varney, J., and Wilkins, M. D. Comparison of two methods for assessing the removal of total organisms and pathogens from the skin. *J. Hyg.* 75:259 (1975).

20. Noble, W. C. Staphylococcus on the hair. *J. Clin. Pathol.* 19:570 (1966).

21. Summers, M. M., Lynch, P. F., and Black, T. Hair as a reservoir for staphylococci. *J. Clin. Pathol.* 18:13–15 (1965).

22. Noble, W. C., and Midgley, G. Scalp carriage of *Pityrosporum* species: The effect of physiology, maturity, sex, race. *Sabouraudia* 16:229 (1978).

23. Marples, R. R., and Towers, A. G. Lab model for investigation of contact transfer of organisms. *J. Hyg.* 82:234 (1979).

24. McBride, M. E., Duncan, W. C., and Knox, J. M. Correlations between epithelial cells and bacterial populations in bacteriologic skin samples. *Br. J. Dermatol.* 99:537 (1978).

25. Pohle, W. D., and Stuart, L. S. The germicidal action of cleansing agents: A study of a modification of Price's procedure. *J. Infect. Dis.* 67:275 (1940).

26. Wilson, P. E. A comparison of methods for assessing the value of antibacterial soaps. *J. Appl. Bacteriol.* 33:574 (1970).

27. Michaud, R. N., McGrath, M. B., and Goss, W. A. Improved experimental model for measuring skin degerming activity on the human hand. *Antimicrob. Agents Chemother.* 2:8 (1972).

28. Burtenshaw, J. M. L. The mortality of *Streptococcus* on the skin and on other surfaces. *J. Hyg.* 38:575 (1938).

29. Williamson, P., and Kligman, A. M. A new method for the quantitative investigation of cutaneous bacteria. *J. Invest. Dermatol.* 45:498 (1965).

30. Bloom, E. Quantitation of skin bacteria: Lethality of the wash solution to remove bacteria. *Acta Dermatovener. (Stockh.)* 59:460 (1979).

31. Kishishita, M., Ozaki, Y., Ushijuma, T., and Ito, Y. New medium for isolation of Propionibacteria and its application to assay the normal flora of facial skin. *Appl. Environ. Microbiol.* 40:1100 (1980).

32. Marples, R. R., Fulton, J. E., Leyden, J., and McGinley, K. J. Effects of antibiotics on the nasal flora of acne patients. *Arch. Dermatol.* 99:647 (1969).

33. Bibel, D. J., and Lovell, D. J. Skin flora maps. A tool in the study of cutaneous ecology. *J. Invest. Dermatol.* 67:265 (1978).

34. Stringer, J. F., and Marples, R. R. Ultrasonic methods for sampling human skin microorganisms. *Br. J. Dermatol.* 94:551 (1978).

35. Thran, V. Mikrobiologische Untersuchung von Oberflachen ein Probennahmegerat. *Fleishwirtschaft* 59:950 (1979).

36. Hartmann, A. A. A comparative investigation of methods for sampling skin flora. *Arch. Dermatol. Res.* 274(3−4):381 (1983).

37. Holland, K. T., Roberts, C. D., Cunliffe, K. T., and Williams, C. D. A technique for sampling microorganisms from pilosebaceous ducts. *J. Appl. Bacteriol.* 37:286 (1974).
38. Hucker, G. H., Emery, A. J., and Winkel, E. Adherence of film to plastics and china surfaces. *J. Milk Food Technol.* 14:95 (1951).
39. Anderson, K. F. Antibacterial bacteriological swabs (Letter). *Br. Med. J.* 2(2):1123 (1965).
40. Douglas, J. Recovery of known numbers of microorganisms from surfaces by swabbing. *Lab. Pract.* 17:1336 (1971).
41. Shaw, C. M., Smith, J. A., McBride, M. E., and Duncan, W. C. An evaluation of techniques for sampling skin flora. *J. Invest. Dermatol.* 54:160 (1970).
42. Rubbo, S. D., and Benjamin, M. Some observations on survival of pathogenic bacteria on cotton wool swabs. Development of a new type of swab. *Br. Med. J.* 1:983 (1951).
43. Evans, C.A., and Stevens, R. J. Differential quantitation of surface and subsurface bacteria of normal skin by the use of cotton swab and scrub methods. *J. Clin. Microbiol.* 3(6):576 (1976).
44. Higgins, M. A comparison of the recovery rage of organisms from cotton wool and calcium alginate wool swabs. *Monthly Bull. Ministry Health Public Lab. Serv.* 9:50 (1950).
45. Notermans, S., Hindle, V., and Kampelmacher, E. H. Comparison of cotton swab versus alginate swab sampling method in the bacteriological examination of broiler chickens. *J. Hyg.* 77:205 (1976).
46. Evans, C. A., and Stevens, R. J. Differential quantitation of surface and subsurface bacteria of normal skin by the use of cotton swabs and scrub methods. *J. Clin. Microbiol.* 3(6):576 (1976).
47. Lovell, L. D. Skin bacteria: Their location with reference to skin sterilization. *Surg. Gynecol. Obstet.* 80:174 (1945).
48. Baxby, D., and Woodruff, R. C. S. The location of bacteria in skin. *J. Appl. Bacteriol.* 28:316 (1965).
49. Somerville, D. A., and Murphy, C. T. Quantitation of *Corynebacterium acnes* on healthy human skin. *J. Invest. Dermatol.* 60:231 (1973).

50. Kooyman, D. J., and Simons, R. W. "Sticky disc" sampling of skin microflora. *Arch. Dermatol.* 92:581 (1965).

51. Malcolm, S. A., and Hughes, T. C. The demonstration of bacteria on and within the stratum corneum using scanning electron microscopy. *Br. J. Dermatol.* 102:267 (1980).

52. Marks, R., and Dawber, R. P. R. Skin surface biopsy: An improved technique for the examination of the horny layer. *Br. J. Dermatol.* 84:117 (1971).

53. Marks, R., and Dawber, R. P. R. In situ microbiology of the stratum corneum. *Arch. Dermatol.* 105:216 (1972).

54. Whiteside, J. A., and Voss, J. S. Incidence and lipolytic activity of *P. acnes* and *P. granulosum* in acne and normal skin. *J. Invest. Dermatol.* 60:94 (1973).

55. Marples, R. R., Leyden, J. J., Stewart, R. N., Mill, O. H., and Kligman, A. M. The skin microflora in acne vulgaris. *J. Invest. Dermatol.* 62:37 (1974).

56. Puhvel, S. M., Reisner, R. M., and Amirian, D. A. Quantification of bacteria in isolated pilosebaceous follicles in normal skin. *J. Invest. Dermatol.* 65:525 (1975).

57. Mustakallio, K. K., Salo, O. P., Kustala, R., and Kustala, U. Counting the number of aerobic bacteria in full thickness suction blisters. *Acta Pathol. Microbiol. Scand.* 69:477 (1967).

58. Wolf, J. Die innere Struktur der Zellen des Stratum desquamans der menochlichen epidermis. *Z. Mikr. Anat. Forsch.* 46:170 (1939).

59. Wolf, J. Das oberflachenrelief der menschlichen Haut. *Z. Mikr. Anat. Forsch.* 47:351 (1940).

60. Pinkus, H. Examination of the epidermis by the strip method of removing horny layers. I. Observations of thickness and on mitotic activity after stripping of the horny layer. *J. Invest. Dermatol.* 16:383 (1951).

61. Milne, L. J. R., and Barnetson, R. St. C. Diagnosis of dermatophytes using vinyl adhesive tape. *Sabouraudia* 12:162 (1974).

62. Weigand, D. A., and Gaylor, J. R. Removal of stratum corneum invivo: An improvement on the cellophane tape stripping technique. *J. Invest. Dermatol.* 43:129 (1964).

63. Updegraff, D. M. A cultural method of quantitatively studying the microorganisms of the skin. *J. Invest. Dermatol.* 43:129 (1964).
64. Houghton, R. H., and May, J. W. Bacteriostatis of *S. aureus* by a volatile component of Scotch brand cellulose adhesive tape. *Nature* 201:1346 (1964).
65. Eade, G. C. The relationship between granulation tissue, bacteria and skin grafts in burned patients. *Plast. Reconstruct. Surg.* 22:42 (1958).
66. Bretano, L., and Gravnens, D. L. A method for the quantitation of bacteria in burn wounds. *Appl. Microbiol.* 15:670 (1967).
67. Georgiade, N. G., Lucas, M. C., O'Fallon, W. M., and Osterhout, S. A comparison of the methods for the quantitation of bacteria in burn wounds. I. Experimental evaluation. *Am. J. Clin. Pathol.* 53:35 (1970).
68. Georgiade, N. G., Lucas, M. C., and Osterhout, S. A comparison of methods for quantitation of bacteria in burn wounds. II. Clinical evaluation. *Am. J. Clin. Pathol.* 53:40 (1970).
69. Saymen, D. G., Nathan, P., Holder, I. A., Hill, E. D., and MacMillian, B. G. Infected wound surface: An experimental model for the quantitation of bacteria in infected tissue. *Appl. Microbiol.* 23:509 (1971).
70. Dutz, W., and Kohout, E. Dermatologic diagnosis by using the hemocytometer and the dental broach. *Int. J. Dermatol.* 21(7):410 (1982).
71. Duguid, J. P., and Wallace, A. T. Air infection with dust liberated from clothing. *Lancet* 2:845 (1948).
72. Lidwell, O. M., Mackintosh, C. A., and Towers, A. G. The evaluation of fabrics in relation to their use as protective garments in nursing and surgery. Dispersal of skin organisms in a test chamber. *J. Hyg.* 81:453 (1978).
73. Mackintosh, C. A., Lidwell, O. M., Towers, A. G., and Marples, M. M. The dimensions of skin fragments dispersed into the air during activity. *J. Hyg.* 81:471 (1978).
74. Noble, W. C. Sampling airborne microbes – Handling the catch. In: *Airborne Microbes, Seventeenth Symposium of*

the *Society for General Microbiology* (P. H. Gregory and J. L. Monteith, eds.). Cambridge University Press, London (1967).

75. Noble, W. C., and Davies, R. R. Studies on the dispersal of staphylococci. *J. Clin. Pathol.* 18:16 (1965).

76. Selwyn, S., and Ellis, H. Skin bacteria and skin disinfection reconsidered. *Br. Med. J.* 1:136 (1972).

9
Skin Permeation: In Vitro Techniques

ROBERT L. BRONAUGH *Division of Toxicological Studies, Food and Drug Administration, Washington, D.C.*

INTRODUCTION

The absorption of chemicals through the skin is now recognized as an important route of entry into the body. The skin is no longer considered simply an impervious protective barrier. Much progress has been made in the understanding of mechanisms of skin permeation in the last 30 years. Percutaneous absorption has been shown to occur by a passive diffusion process that takes place with all chemicals to varying degrees depending on their physicochemical properties.

The stratum corneum has been shown to be the primary barrier to the absorption of water (1) and certain alkanols (2); and by extrapolation of these findings, it is considered to be the primary barrier to absorption of other water-soluble compounds. For hydrophobic compounds, the aqueous viable epidermal and dermal tissues are significant barriers to systemic absorption (3). Compounds must diffuse about 200 µm to reach the blood vessels located in the papillary dermis.

Because absorption through the skin is a passive diffusion process, similar values for the penetration of chemicals through skin can be obtained with either in vivo or in vitro techniques if proper methods are used (4-6). The rate of permeation through the skin can be measured more accurately using in vitro methods, in which samples are obtained directly beneath the skin in the diffusion cell. A small number of highly reactive compounds may be metabolized in skin to a significant extent. It has yet to be demonstrated that metabolism will affect the rate of permeation of a chemical through skin.

TYPES OF DIFFUSION CELLS

Many types of diffusion cell systems have been used for the measurement of percutaneous absorption. Sometimes the nature of the study will dictate the use of a particular type of cell. In many cases, however, the personal preference of the ivnestigator will decide this issue.

For the study of the steady-state diffusion of molecules in solution across a membrane, the two-chambered (side-by-side) cell is popular. Rates of diffusion can be accurately measured as solutions on both sides of the membrane are continually stirred. The cell is useful for measuring the absorption of compounds from aqueous solutions under steady-state conditions that occurs with the application of large (infinite) doses. A practical use is in the determination of the expected steady-state permeation rates of a compound applied to skin in a drug delivery device.

An approach that more closely duplicates usual exposure or product use conditions is the application of a finite dose in a thin layer to the surface of the skin (7) (see discussion of finite and infinite dosing in "Application of Compound"). A decrease in surface concentration of a compound occurs as the material is absorbed, and therefore a steady-state rate is not obtained. A thin vehicle layer can be applied to skin in a one-chambered cell. A receptor compartment beneath the skin, which contains saline or a physiological buffer that is mixed with a stirring bar, collects absorbed material. The compound can be applied to skin in any desired vehicle. The surface of the skin does not become overly hydrated as a result of continued contact with water as it does in the two-chambered cell.

The flow-through cell (8) has several advantages compared to the one-chambered cell. The receptor fluid is pumped beneath the surface of the skin, continually removing absorbed material and maintaining sink conditions. Samples can be collected automatically in a fraction collector 24 hr/day. The volume of the receptor must be small so that a mangeable volume of receptor fluid is obtained. Bronaugh and Stewart (8) used a flow rate of at least 1.5 ml/hr in their cell, which had a receptor volume of 0.13 ml. The flow cell is preferable for use in experiments where continual replacement of the receptor fluid is desired. For maintenance of skin viability, an organ culture medium can be continuously pumped beneath the skin (9).

For some chemicals (i.e., mosquito repellent, fragrances), evaporation from the skin is limited to produce the desired effect, and percutaneous absorption is minimized. Diffusion cells have been designed that, in addition to measuring absorption through the skin, trap the material evaporating from the skin surface so that it can be analyzed (10,11). Spencer and co-workers (10) used collecting tubes (containing cotton as the absorbent) above the skin surface to trap evaporating diethyltoluamide. Alternatively, the top of the cell could be connected to a bubbler trap containing counting solution so that air passing over the skin carries evaporating radiolabeled material to the trap. Reifenrath and Robinson (11) have described a flow-through cell for the quantitation of absorption and evaporation. The evaporating compound is drawn by a stream of air through a vapor trap containing absorbent material; it then goes into a bubbler trap containing scintillation counting solution that serves as a safety trap.

CHOICE OF SKIN

Human skin is preferred for the most accurate work but care must be taken to be sure that the barrier is intact (see "Variability of Results"). For the study of a toxic chemical, in which in vivo measurements would be precluded, in vitro methodology is the only way to examine absorption through human skin. In general, the use of animal skin leads to an overestimation of human percutaneous absorption. For many studies, however, it is best to conduct experiments with an animal skin that is in plentiful supply and then to verify the conclusions with human skin.

The rat is a convenient animal for these studies, and absorption values obtained with rat skin are frequently in reasonable agreement with human data (12) (Fig. 1). The permeability of hairless mouse skin was shown to be similar to that of human skin

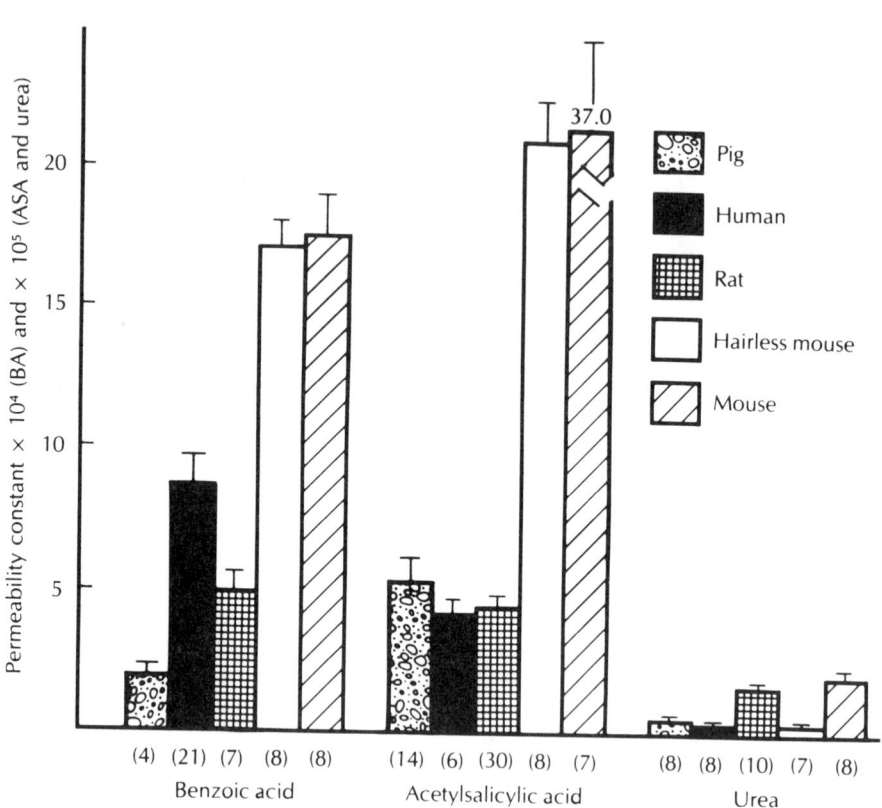

Figure 1 Permeability constants obtained with human and animal skin. Values are the mean ± SE of the number of determinations in parentheses. BA = benzoic acid; ASA = acetylsalicylic acid. (From Ref. 12, reproduced with permission of the publisher.)

for a series of straight-chain alkanols (13) and certain corticosteroids (14), but for other compounds absorption was faster through human skin (12,15,16).

PREPARATION OF SKIN

Skin should be prepared for use in the diffusion cell to give a barrier that is similar to the in vivo barrier layer. The microcirculation is located in the papillary dermis, approximately 200 µm below the surface of the skin (17). Compounds absorbed through the skin would have to diffuse this distance in order to be taken up by the blood that perfuses the skin. A layer of skin approximately 200 µm thick is therefore similar to the in vivo barrier. This layer consists of the whole epidermis and the papillary dermis where the capillary loops are found.

A dermatome is convenient for preparing skin for absorption studies (18). For human skin and the skin of the fuzzy rat, a 200-µm section of skin can easily be prepared. When preparing the skin of the haired rat, a 350-µm section is as thin as can be made without some damage to the barrier properties of the skin. The reason for this damage is not known but is likely due to disruption of the stratum corneum at the location of the hair follicles as the blade passes through the area.

The epidermal layer has been used as the absorption barrier in studies with human skin that is free of hair. The epidermis is readily separated from full-thickness nonhairy skin by submerging full-thickness skin in 60°C water for 1 min. The epidermal layer can then be peeled away with forceps. The viability of the skin is destroyed during the separation process and therefore pharmacokinetic studies cannot be carried out.

In general, full-thickness skin should not be used for in vitro studies because it presents a barrier layer that is much thicker than that traversed by a diffusing molecule in the in vivo situation. This is particularly important for hydrophobic compounds that do not readily penetrate the aqueous dermal tissue (3). When using animals with thin skin, such as the rabbit and mouse (thickness less than 1 mm), the preparation of split-thickness skin may be difficult and is probably not required.

RECEPTOR FLUID

The choice of the receptor fluid beneath the skin can greatly influence the experimental results. The fluid must allow free diffusion of the compound into it from the skin if sampling of the receptor fluid is to be used as a measure of percutaneous absorption. The receptor fluid must not alter the integrity of the barrier layer. For compounds soluble in water, a normal saline or aqueous buffer solution is often used. A tissue culture medium can be used for pharmacokinetic studies.

Recently we have demonstrated the importance of the receptor fluid in the evaluation of the absorption of hydrophobic compounds (3). Compounds that are soluble in oil but not in water are often absorbed readily into the skin but will not partition freely out of the skin into an aqueous receptor fluid. The permeation of two water-insoluble fragrance ingredients, cinnamyl anthranilate and acetyl ethyl tetramethyl tetralin (AETT), was examined. The enhancement of the absorption of these compounds by different receptor fluids was determined. Any damage to the skin with the use of a receptor fluid was detected by the simultaneous measurement of a control compound (cortisone). If cortisone absorption increased above normal during the course of the experiments, it could be assumed that increased absorption of the fragrance was due to a damaged skin barrier. The greatest absorption of cinnamyl anthranilate without apparent damage to the skin occurred with the use of the nonionic surfactant PEG-20 oleyl ether (Volpo 20; Table 1). The in vitro permeation of the compound was more than 60% of that determined by in vivo methods. Other nonionic surfactants resulted in either less absorption of cinnamyl anthranilate or damage to the skin. Neither rabbit serum nor bovine serum albumin was comparable to the 6% solution of Volpo 20. Similar results were obtained when the permeation of AETT was measured in the different receptor fluids.

Best results with animal skin are obtained with the use of sparsely haired animals (18). This allows the preparation of thinner (200 µm), more physiological sections of skin with a dermatome. As previously mentioned, it is particularly important to use a skin preparation with minimal dermal tissue when studying hydrophobic compounds. With the thinner barrier layer, more dilute solutions (0.5%) of Volpo 20 could be used and in vitro and in vivo absorptions were almost identical.

The values in Table 2 can be used as guidelines to help determine when the solubility properties of a test compound require that a lipophilic receptor fluid be used. The ratio of the absorption values obtained with Volpo 20 and normal saline as the receptor fluid increases as the water solubility of the compound decreases. Enhanced absorption with the surfactant solution apparently can be expected when the water solubility of a compound is in the milligrams/liter range and when the compound is more soluble in a lipoidal solution.

It is possible to maintain viability of skin in a flow-through diffusion cell using a tissue culture medium as the receptor fluid (9). The value of this approach in measuring percutaneous absorption is as yet unclear. For most compounds it seems unlikely that metabolism would play a role in determining the rate of absorption through the skin. With compounds that are metabolized to potent, pharmacologically active compounds in the skin, maintenance of skin viability may be of interest. Care must be taken

Table 1 Effect of Diffusion Cell Conditions on the Absorption of Cinnamyl Anthranilate (Cortisone Control)[a]

Receptor fluid	Skin preparation	Cinnamyl anthranilate (% absorbed in 5 days)	Cortisone permeability constant × 10^5
Normal saline (4)	Whole skin	5.4 ± 0.3	3.8 ± 0.7
1.5% Volpo 20 (4)	Whole skin	5.4 ± 0.9	—
Normal saline (4)	350 μm	5.8 ± 0.4	7.1 ± 0.5
1.5% Volpo 20 (10)	350 μm	15.5 ± 1.2	6.1 ± 0.5
6% Volpo 20 (8)	350 μm	27.9 ± 1.8	7.0 ± 0.9
20% Volpo 20 (8)	350 μm	18.3 ± 1.8	9.3 ± 0.9
Rabbit serum (4)	350 μm	8.8 ± 0.6	6.8 ± 0.8
3% bovine serum albumin (4)	350 μm	12.1 ± 1.2	5.4 ± 0.2
50:50 methanol: water (4)	350 μm	27.1 ± 2.0	17.2 ± 0.2
1.5% Triton-X (4)	350 μm	17.9 ± 1.1	10.8 ± 0.5
6% Triton-X (4)	350 μm	38.4 ± 2.9	14.5 ± 1.3
6% Pluronic F68 (4)	350 μm	7.3 ± 1.3	9.8 ± 0.6

[a]Values are the mean ± SE of the number of determinations in parentheses. For most experiments, a 350-μm section from the surface of whole rat skin was prepared with a Padgett Electrodermatome. Compounds were applied to skin in a petrolatum vehicle. In vivo absorption of cinnamyl anthranilate was 45.6%.

so that bacterial growth on the skin or in the medium does not result in bacterial metabolism that interferes with the observation of metabolism by skin enzymes.

The receptor fluid is often continuously mixed during an experiment. With flow-through cells, the movement of the receptor fluid through the small receptor volume provides the mixing. Sink

Table 2 Comparison of Solubility Properties and the Effect of Volpo 20 on Percutaneous Absorption

Compound	Water solubility (mg/liter)	Octanol/ water (K)	Skin permeation ratio[a] 6% Volpo 20/ saline
Urea	1 × 16⁶	0.002	1.1[b]
Cortisone	280	44	1.2[c]
Testosterone	11	2089	2.3[d]
Cinnamyl anthranilate	0.23	652	3.1[c]
AETT	0.012	3589	30.0[e]

[a]The skin permeability ratio was determined by comparing the amount of compound absorbed in experiments using the two receptor fluids. Specific conditions for each compound (vehicle-length of experiment) were: [b]water vehicle—43 hr; [c]acetone vehicle—5 days; [d]acetone vehicle—43 hr; [e]petrolatum vehicle—5 days.
Source: Ref. 3, reproduced with permission of the publisher.

conditions must be maintained so that a free partitioning of test compound from skin into the receptor fluid occurs. When radioisotopes are used, the small amount of compound applied to the skin does not usually create a problem. If large amounts of material are applied to the skin, high concentrations of material in the receptor fluid should be avoided. Possible solutions would include frequent changing of the receptor fluid, use of flow-through cells, or use of static cells with large receptor volumes.

APPLICATION OF COMPOUND

It is frequently desirable to apply compounds to the skin to simulate exposure or "use" conditions. When exposure to a chemical may occur from more than a single vehicle, testing with vehicles of varying solubility properties is suggested. Examples of commonly used vehicles in different classes are: polar vehicle—

water or propylene glycol; lipoidal vehicle—petrolatum or isopropyl myristate (19). A vehicle dose of 5 mg/cm^2 skin approximates the amount applied under normal use. When a volatile vehicle such as water is used, occlusion of the site of application is necessary if a constant concentration of test compound is to be maintained.

The amount of material applied to skin is frequently insufficient to maintain a steady-state rate of absorption. The absorption rate increases to a maximum and then immediately decreases as the concentration of material on the surface of the skin decreases. The amount applied is therefore called a finite dose.

When the chemical in contact with the skin is more concentrated, the relatively small amount that is absorbed does not significantly change the surface concentration. The amount of material applied can be referred to as an infinite dose and results in the eventual attainment of a steady-state rate of percutaneous absorption. If the concentration on the surface of the skin is known, a permeability constant can be calculated. The dose applied in a drug delivery device is often an infinite dose since sustained absorption is desirable.

The duration of an absorption study depends on many factors. Frequently we have measured the amount of test compound absorbed after 24 hr of skin contact. In in vivo experiments, material applied to the skin that is unabsorbed at 24 hr begins to be sloughed off as a result of the turnover of stratum corneum cells. For in vitro studies of greater than 24-hr duration, it may be advantageous to wash the surface of the skin at 24 hr so that results will be more comparable to the in vivo situation.

CALCULATION OF DATA

The method of expressing absorption data is usually determined, at least in part, by the type of diffusion cell study conducted. With finite dosing of a compound, a permeability constant cannot be calculated because steady-state absorption is not achieved. This constant cannot be determined when a volatile vehicle is used because the concentration of compound on the surface of the skin is unknown after the vehicle evaporates. The amount of material that penetrates skin in these circumstances is often expressed in terms of the amount applied (i.e., the percent of the applied dose absorbed). The maximum rate and a graph of the time course of absorption are helpful additional information.

If an infinite dose is applied and the surface concentration is known, a permeability constant can be calculated by dividing the steady-state absorption rate by the applied concentration. A permeability constant can be particularly useful when the absorption of compounds is compared since absorption is normalized for differences in concentration. The linear relationship between dose and absorption may only hold at relatively low concentrations.

VARIABILITY OF RESULTS

When human cadaver skin is used as the membrane, concerns about variability are most pronounced (20). Because of our heterogeneous population, human skin can differ in its permeability properties as much as fivefold. Also, the barrier properties of skin from cadavers may be altered. Many variables may influence the condition of this skin, including time from death to skin harvest, scrubbing or cleansing of the skin with antibacterial agents, and storage of the skin from harvest until use. It is therefore advisable to pretest cadaver skin with a standard compound with known permeability properties. Tritiated water has been useful for this purpose (20). The average permeability constant for 43 donors was 1.5×10^{-3} cm/hr. Changes in water permeability were reflected also in the permeability of seven compounds with varying permeability/solubility properties (Fig. 2). Although water is a small, polar molecule, conditions in skin that increase its permeability also seem to increase the permeability of more lipophilic compounds.

Within a given strain of animals there is less variability, but it cannot be ignored. Skin freshly removed from an animal does not need to be pretested with a standard compound, but experiments should be conducted on more than one animal and the values averaged to minimize variability.

CONCLUSIONS

Accurate percutaneous absorption results can be obtained using in vitro procedures. Care must be taken to avoid the many potential problems discussed in this chapter. The specific method of choice will often depend on the type of compound and the goals of the investigation.

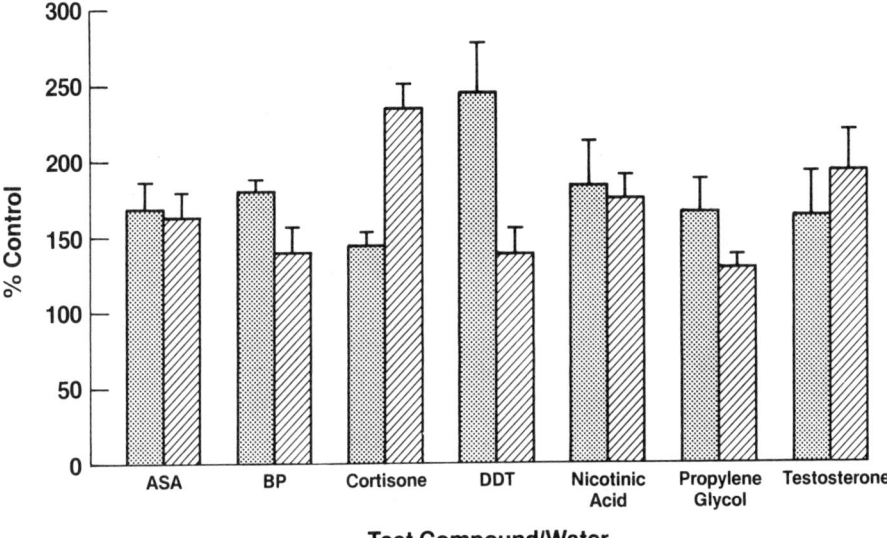

Figure 2 Absorption of water as an indicator of permeability changes. (▒) test compound; (▨) water. Values are the mean ± SE of five determinations. Absorption of each test compound and water was measured in two different pieces of skin. Control values are the absorption values for the test compound and water through the skin sample with the lowest water permeability. The increase in water permeation was significantly different (t test, $p < 0.05$) from the increase in test compound permeation for two compounds, cortisone and DDT. ASA = acetylsalicylic; BP = benzo(a)pyrene. (From Ref. 20, reproduced with permission of the publisher.)

REFERENCES

1. Berenson, G. S., and Burch, G. E. Studies of diffusion through dead human skin. Am. J. Trop. Med. Hyg. 31:842–853 (1951).

2. Scheuplein, R. J., and Blank, H. I. Permeability of skin. Physiol. Rev. 51:702–747 (1971).

3. Bronaugh, R. L., and Stewart, R. F. Methods for in vitro percutaneous absorption studies. III. Hydrophobic compounds. J. Pharm. Sci. 73:1255–1258 (1984).

4. Franz, T. J. Percutaneous absorption. On the relevance of in vitro data. J. Invest. Dermatol. 64:190–195 (1975).

5. Bronaugh, R. L., Stewart, R. F., Congdon, E. R., and Giles, A. L., Jr. Methods for in vitro percutaneous absorption studies. I. Comparison with in vivo results. Toxicol. Appl. Pharmacol. 62:474–480 (1982).

6. Bronaugh, R. L., and Maibach, H. I. Percutaneous absorption of nitroaromatic compounds: In vivo and in vitro studies in the human and monkey. J. Invest. Dermatol. 84:180–183 (1985).

7. Franz, T. J. The finite dose technique as a valid in vitro model for the study of percutaneous absorption in man. Curr. Probl. Dermatol. 7:58–68 (1978).

8. Bronaugh, R. L., and Stewart, R. F. Methods for in vitro percutaneous absorption studies. IV. The flow-through diffusion cell. J. Pharm. Sci. 74:64–67 (1985).

9. Holland, J. M., Kao, J. Y., and Whitaker, M. J. A multisample apparatus for kinetic evaluation of skin penetration in vitro. Toxicol. Appl. Pharmacol. 72:272–280 (1984).

10. Spencer, T. S., Hill, J. A., Feldmann, R. J., and Maibach, H. I. Evaporation of diethyltoluamide from human skin in vivo and in vitro. J. Invest. Dermatol. 72:317–319 (1979).

11. Reifenrath, W. G., and Robinson, P. B. In vitro skin evaporation and penetration characteristics of mosquito repellents. J. Pharm. Sci. 71:1014–1018 (1982).

12. Bronaugh, R. L., Stewart, R. F., and Congdon, E. R. Methods for in vitro percutaneous absorption studies. II. Animal models for human skin. Toxicol. Appl. Pharmacol. 62:481–488 (1982).

13. Durrheim, H., Flynn, G. L., Higuchi, W. I., and Behl, C. R. Permeation of hairless mouse skin. I. Experimental methods and comparison with human epidermal permeation by alkanols. J. Pharm. Sci. 69:781–786 (1980).

14. Stoughton, R. B. Animal models for in vitro percutaneous absorption. In: Animal Models in Dermatology (H. I.

Maibach, ed.). Churchill-Livingstone, New York, pp. 121–132 (1975).

15. Walker, M., Dugard, P. H., and Scott, R. C. In vitro percutaneous absorption studies: A comparison of human and laboratory species. *Hum. Toxicol.* 2:561–562 (1983).

16. Behl, C. R., Flynn, G. L., Linn, E. E., and Smith, W. M. Percutaneous absorption of corticosteroids: Age, site, and skin sectioning influences on rates of permeation of hairless mouse skin by hydrocortisone. *J. Pharm. Sci.* 73:1287–1290 (1984).

17. Barry, B. W. Skin structure, function, diseases, and treatment. In: *Dermatological Formulations* (B. W. Barry, ed.). Marcel Dekker, New York, pp. 1–48 (1983).

18. Bronaugh, R. L., and Stewart, R. F. Methods for in vitro percutaneous absorption studies. VI. Preparation of the barrier layer. *J. Pharm. Sci.* 75:487–491 (1986).

19. Bronaugh, R. L., Congdon, E. R., and Scheuplein, R. J. The effect of cosmetic vehicles on the penetration of N-nitrosodiethanolamine through excised human skin. *J. Invest. Dermatol.* 76:94–96 (1981).

20. Bronaugh, R. L., Stewart, R. F., and Simon, M. Methods for in vitro percutaneous absorption studies. VII. Use of excised human skin. *J. Pharm. Sci.* 75:1094–1097 (1986).

10
The Evaluation of Acne in Humans

ROBERT L. RIETSCHEL *Ochsner Clinic, New Orleans, Louisiana*

DIRECT METHODS

A variety of methods have been used to evaluate acne in humans. Direct methods include lesion counts and global grading systems. Indirect methods include measuring sebum production, lipid composition, comedolysis, and microflora quantification. The direct methods assess efficacy of treatment and allow limited inferences regarding the mechanism of action of the study agent. Indirect methods collect data that provide answers to questions of mechanism while giving limited inferences as to efficacy.

Lesion Counts vs. Global Scoring

Lesion counts are attractive because of the exact quantification produced. In published reports one reads exact numbers of papules, pustules, and comedones but seldom are the definitions given. Was a 0.5-cm, palpable, inflamed lesion without obvious blackhead determined to be an inflamed closed comedo, a papule, a nodule, or a cyst? Would any two observers have reached the same count or conclusion? The matter has received some scrutiny (1). Multiple observers grading the same patients tend to grade papules more concordantly than either pustules or comedones (1). Suggested definitions for lesion counting are found in Table 1, but an alternative set of definitions with greater detail can be found in Cunliffe and Cotterill (2). Because papules and comedones can be numerous and tedious to count on the full face, one half of the face may be chosen and counted from mandible to hairline. The paucity of pustules and nodulocystic lesions makes it both practical and desirable to count both sides of the face even when only half is being followed for comedones and papules. Virtually no one has addressed the issue of counting the macular erythema that may persist as lesion activity fades. This author does not count macular erythema as it represents resolved disease activity. However, by combining lesion counts with global assessments a clearer picture of the magnitude of improvement can be achieved. Ideally, the two grading systems would produce similar results. This has proved to be the case

Table 1 Suggested Definitions for Lesion Counting

Lesion	Suggested definition
Comedo	Blackhead or horny plug in which follicular obstruction is present with inflammation absent or less than 2 mm in diameter
Papule	Inflammation predominates over follucular obstruction which is usually inapparent or dominated by 2 mm of inflammation; may extend to 5–6 mm in diameter
Nodulocystic lesions	Inflammatory papules with deeper dermal inflammation and greater than 6 mm in diameter; cysts are more fluctuant than nodules
Pustules	Superficial whitish-yellow lesions with or without an erythematous base; generally not larger than 3 mm in diameter

(3). Independent global scales for individual lesion types are equally reproducible to exact counts and much easier to use (3).

Multiple observers actually produced highly concordant global grades (1,3) using an eight-point scale devised by Cook et al. (4) when grading was performed live. Observers who, instead of grading subjects in person, reviewed photographs of acne subjects at a later date and applied a similar global scale tended to score lower grades for the most active acne and more concordant grades for less active acne (5). Thus, global grading systems have equal reproducibility to lesion counts and offer the advantage of speed. However, experience improves concordance, making it incumbent on new researchers to validate their methodology by comparing both systems in their hands (3).

Time Intervals

The frequency of reevaluation of acne, whether evaluated by lesion counts, global assessment, or both, generally is at 2- or 4-week intervals. The 2-week interval is most useful when a difference in speed of improvement is anticipated; otherwise, a 4-week interval is adequate. Duration of evaluation should extend 10—12 weeks to achieve differences that are evaluable by lesion count or global grading systems and allow for epidermal turnover (28 days) to have passed through several cycles. Antiandrogen treatments may require additional months to be fully evaluated, especially in women, to allow several menstrual cycles (6).

Sebum Excretion Rates

Sebum production is under androgenic control. Excess circulating androgens are not necessary for increased sebum production, for enhanced sensitivity of the sebaceous glands to normal levels of hormone may likewise lead to excess sebum. Excess circulating androgens have frequently been identified in women with acne (7). Classes of agents that lower sebum production include retinoids (8), estrogens (9), antiandrogens (6), corticosteroids (10), and superficial X-ray therapy (11). Methods of determining sebum excretion rates have undergone recent improvements in methodology as reported by Harris et al. (12).

Lipid Composition

Free fatty acids and the ratio of free fatty acids in sebum to triglycerides have been of interest because of the irritancy and comedogenicity of this component of sebum and its seeming relationship to bacterial degradation of triglycerides into free fatty acids. However, this subject is controversial because authors have variously reported increased or decreased levels of free fatty acids, triglycerides, and wax esters. This has been reviewed nicely by Cunliffe and Cotterill (2). Squalene is generally increased and has received renewed interest recently (13), as has linoleic acid (14). A 95% reduction of *Propionibacterium acnes* organisms leads to a 40-50% reduction in free fatty acids, which has caused some authors to use free fatty acid levels as an indirect reflection of antimicrobial activity (15). Yet it must be remembered that direct inhibition of lipase may occur with antibiotics (especially tetracycline) without antimicrobial action and lead to lower free fatty acid levels without bactericidal action (15). This author finds lipid composition the most indirect monitor of the acne activity, and its use is restricted to individuals with predetermined high levels prior to study. This has been the practice in the most encouraging reports on the value of lipid composition determination (16).

Microflora

The concentration of surface and follicular bacteria can be determined by the methods outlined in the earlier chapter on cutaneous bacteriology. Surface methods are most commonly employed because of the ease with which samples can be taken. Intrafollicular activity is likely to be of greater significance, and measurement of this area would be better than simple surface sampling.

Comedolysis

The degree of follicular impaction with keratinaceous debris has been quantitated by placing a drop of cyanoacrylate glue on the area to be studied and pressing a glass slide over the area, allowing the glue to dry (20-40 sec), and then removing the slide (17). This procedure can be as unpleasant as any epilation technique. A new sialastic impression technique may provide similar information with less discomfort (18).

CONCLUSION

The combination of parameters chosen for studying acne depends on the objective. If clinical improvement is of primary interest, lesion counts combined with global scores are more than adequate. Using both decreases the likelihood of criticism until such time as one's own data have validated equivalency in one's own hands. The addition of surface comedo assay by cyanoacrylate glue or sialastic imprint provides good intercenter standardization for multicenter studies, if there is some question of the standardization of grading scales. Surface and intrafollicular microbiological quantification (predominantly of P. acnes) and sebum excretion rates tend to tell more about mechanism of action than does lipid fractionation at this time. If recent theories regarding squalene or linoleate prove correct, a direct measure of these fractions would be worthwhile.

REFERENCES

1. Feucht, C. L., Allen, B. S., Chalker, D. K., and Smith, J. G., Jr. Topical erythromycin with zinc in acne. *J. Am. Acad. Dermatol.* 3:483–491 (1980).

2. Cunliffe, W. J., and Cotterill, J. A. *The Acnes*. Saunders, Philadelphia (1975).

3. Allen, B. S., and Smith, J. G. Various parameters for grading acne vulgaris. *Arch. Dermatol.* 118:23–25 (1982).

4. Cook, C. H., Centner, R. L., and Michaels, S. F. An acne grading method using photographic standards. *Arch. Dermatol.* 115:571–575 (1979).

5. Samuelson, J. S. An accurate photographic method for grading acne: Initial use in a double-blind clinical comparison of minocycline and tetracycline. *J. Am. Acad. Dermatol.* 12:461–467 (1985).

6. Miller, J. A., Wojnarowska, F. T., Dowd, P. M., et al. Antiandrogen treatment in women with acne: A controlled trial. *Br. J. Dermatol.* 114:705–716 (1986).

7. Schiavone, F. E., Rietschel, R. L., Sgoutas, D., and Harris, R. Elevated free testosterone levels in women with acne. *Arch. Dermatol.* 119:799–802 (1983).

8. Strauss, J. S., Stanieri, A. M., and Farrell, L. N. The effect of marked inhibition of sebum production with 13-*cis*-rectinoic acid on skin surface lipid composition. *J. Invest. Dermatol.* 74:66–67 (1980).

9. Pochi, P. E., and Strauss, J. S. Sebaceous gland suppression with ethinyl estradiol and diethylstilbesterol. *Arch. Dermatol.* 108:210 (1973).

10. Pochi, P. E., and Strauss, J. S. Effect of prednisone on sebaceous gland secretion. *J. Invest. Dermatol.* 49:456 (1967).

11. Strauss, J. S., and Kligman, A. M. Effect of X-rays on sebaceous glands of the human face: Radiation therapy of acne. *J. Invest. Dermatol.* 33:347 (1960).

12. Harris, H. H., Downing, D. T., Stewart, M. E., and Strauss, J. S. Sustainable rates of sebum secretion in acne patients and matched control subjects. *J. Am. Acad. Dermatol.* 8:200–203 (1983).

13. Saint-Leger, D., Bagne, A., Lefebvre, E., et al. A possible role for squalene in the pathogenesis of acne. II. In vivo study of squalene oxides in skin surface and intra-comedonal lipids of acne patients. *Br. J. Dermatol.* 114:543–552 (1986).

14. Downing, D. T., Stewart, M. E., and Wertz, P. N. Essential fatty acids and acne. *J. Am. Acad. Dermatol.* 14:221–225 (1986).

15. Plewig, G., and Kligman, A. M. *Acne Morphogenesis and Treatment.* Springer-Verlag, New York (1975).

16. Strauss, J. S., and Stranieri, A. M. Acne treatment with topical erythromycin and zinc: Effect on *Propionibacterium acnes* and free fatty acid composition. *J. Am. Acad. Dermatol.* 11:86–89 (1984).

17. Marks, R., and Dawber, R. P. R. Skin surface biopsy, an improved technique for the examination of the horny layer. *Br. J. Dermatol.* 84:117 (1971).

18. Tucker, S. B., Flannigan, S. A., Dunbar, M., Jr., and Drotman, R. B. Development of an objective comedogenicity assay. *Arch. Dermatol.* 22:660–665 (1986).

11
Transepidermal Water Loss: Methods and Applications

THOMAS S. SPENCER *Cygnus Research Corporation, Redwood City, California*

INTRODUCTION

Transepidermal water loss (TEWL) is one mechanism of water loss from the body, distinct from perspiration and exhaled water vapor from the pulmonary system. To be more specific about transepidermal water loss is difficult because individuals will perspire at low levels below that observed as beads of sweat. Low-level perspiration has been described as a constant flux through sweat ducts, a heat pipe phenomenon (1) in addition to diffusion of water vapor through the stratum corneum. For the purposes of the current discussion, transepidermal water loss is defined as insensible water loss through the skin separate and distinct from active perspiration. However, in the methods section, where practical aspects of measuring transepidermal water loss are discussed, perspiration processes, including heat and nervous perspiration, that are not readily apparent are a concern in measurement of transepidermal water loss.

Transepidermal water loss has many applications, including the measurement of water loss from normal skin as well as from skin with various pathological conditions. In general, if the skin is damaged, irritated, or wounded, transepidermal water loss will increase. Other uses of transepidermal water loss include the evaluation of skin moisturization with topically applied toiletry products, loss of aqueous vehicle from the skin surface, and measurement of skin surface water. Wound healing processes can be quantitated with transepidermal water loss, while at the opposite extreme transepidermal water loss can be a reasonably sensitive indicator of the barrier function of skin in permeability studies, both in vivo and in vitro.

In vivo and in vitro with diffusion chambers the dermal side of the skin is exposed to a high concentration of water with an activity approaching one. The mathematics of the diffusion process across the skin under these circumstances can be described from Fick's diffusion equation (2) as follows:

$$J_s = \frac{K_m D}{d} (C_d - C_r) \tag{1}$$

The permeability constant K_p is defined as

$$\frac{J_s}{(C_d - C_r)}$$

Therefore

$$K_p = \frac{K_m D}{d} \tag{2}$$

where

J_s = the steady-state flux of the solute in gm/m²/hr
K_p = the permeability constant (cm s^{-1})
K_m = the partition coefficient
D = the diffusivity of water (cm² s^{-1})
d = thickness of the stratum corneum membrane (cm)
C_d = concentration of water in the donor side (gm/cm^{-3})
C_r = concentration of water at the upper surface of the stratum corneum proportional to the activity of water vapor in the air immediately above the stratum corneum (g/cm³)

The mathematical description of the diffusion of water through the skin presents a number of issues that should be considered when developing protocols to study transepidermal water loss. First, the diffusivity of water in straum corneum is a function of the water content of the stratum corneum. Dry statum corneum tends to have a lower diffusion coefficient than hydrated stratum corneum. The diffusion or water content of the stratum corneum is a gradient from the epidermal side (donor) of stratum corneum to the external environment or air (receptor). Several definitions of this gradient have been described in the literature (3,4); however, our further considerations of measurements of transepidermal water loss assume a constant skin environment resulting from relatively constant ambient humidity. In the case of the donor side, the water activity is assumed to approach one; i.e., there is a constant potential within the body tissues for diffusion of water across the stratum corneum. The activity of water at the skin surface is assumed to be affected by the relative humidity or activity of water in the air above the skin surface. This approximation is valid as long as the air above the skin is not extremely dry and the body temperature is low enough that perspiration is not occurring. Other difficulties include the fact that the thickness of the stratum corneum membrane varies with water content, thereby affecting flux of water across the skin. In subsequent sections the various methods, the utility of those methods, and practical aspects of measuring transepidermal water loss are discussed.

METHODS

Gravimetrics

Gravimetric methods have been used for centuries to evaluate the loss of moisture from the body. The simplest approach is measurement of whole body weight before and after a study to evaluate loss of moisture through the skin; however, loss of moisture by perspiration and pulmonary function is a complicating factor. Gravimetric methods have also been used in which microscopic substances are placed in contact with the skin to evaluate total moisture transport across the skin. The interaction of hygroscopic salt with the skin surface gives rise to the question whether or not these salts will affect the transport of water across the skin. The most practical aspect of gravimetric methodology today is the use of pads to absorb water that is transpiring through the skin. In particular, sweat pads are commonly used to evaluate the efficacy of antiperspirants. The pads are placed under the armpits of the individual before entering a hot chamber and are collected and weighed after a period of exposure. This method is a pragmatic means of identifying the efficacy of antiperspirants on a body site to which antiperspirants are most commonly applied. The sweat pads have the advantage of conforming to the contour of the axilla, being portable, and of being a relatively simple technology for evaluation of perspiration.

Closed Cell Measurements

An extension of the gravimetric method measurements is the closed cell containing a hygroscopic substance that is placed on the skin surface to absorb moisture transpiring through the skin and into the air in a closed chamber (5,6). The small chamber is portable and can be applied to the skin for evaluation after a period of time appropriate to a given study. The method is subject to errors due to saturation of the hygroscopic material or equilibrium effects of the partial vapor pressure of water above the hygroscopic material. Another way of applying the same technique is to use a hygrometer above the skin surface to measure the moisture content of the air above the skin and extrapolating the changing relative humidity to evaluate the water transport across the skin. Evaluation of TEWL using closed cells is limited because lengthy periods of time are required to absorb sufficient moisture for an accurate evaluation of water loss. Holding subjects for that extended period of time is difficult.

Open Cell Methods

A more practical approach than closed cell measurement is the open cell detection of moisture loss through the skin. A stream of air is passed across the skin in which the moisture content of the air stream is measured before and after entering the chamber attached to the skin surface (7-10). A number of detection techniques have been used, including hygrometers, infrared sensors, and gravimetric collection of moisture passing across the skin surface. However, the most common method used has been electrolytic water sensors to detect moisture content just prior to and after passing dry air across the skin. One instrument in this category is the electrolytic water analyzer that has been used extensively by Spruit (11) and Thiele and Malten (12) to evaluate transepidermal water loss. The advantage of the electrolytic detector is that, once calibrated, the sensors provide a reasonably accurate and reproducible assessment of transepidermal water loss. Spruit, Thiele, and Malten have used this device to study transepidermal water loss under a variety of conditions, relative humidities, stress levels, and temperatures (11,12). One major disadvantage of the electrolyte water analyzer is a tendency to become saturated if too much moisture is passed across the sensor in a short period of time. When the sensors become saturated, the calibration of the instrument changes and the response time is very slow. In addition, the electrolytic water analyzers tend to have a slight lag in response time under normal conditions. The success seen by Spruit and Malten can be attributed in large part to their attention to detail in calibration and maintaining calibration of the electrolytic sensors as well as their willingness to replace inaccurate sensors.

In the closed cell flow system, a number of different cell diameters and cell materials have been used. The most common design is a stainless steel hat with inlets and outlets at 45° angles to the skin surface and a cross-section of approximately 1 cm^2. The diameter of the flow cell applied to the skin can affect the measurement of transepidermal water loss because of edge effects, i.e., horizontal diffusion of water from adjacent stratum corneum under the edges of the cell into the measurement chamber. Small-diameter cells have proportionally greater edge effects than large-diameter cells. Typically, dry gas is passed across the skin surface and analyzed for uptake of water on the receiver side. This uptake of water is directly translated into water loss per surface area per unit of time. Difficulties occur when excessive wetness at the skin surface causes condensation in the electrolytic cell. Condensation can be prevented by controlling temperature of flow lines and cells.

The data output from a flow cell just applied to the skin surface is typified by a large peak resulting from surface moisture being extracted from the skin by the dry air stream passed across the skin. After 20–30 min, loss of moisture from the skin surface reaches a steady state. The initial peak is a measure of water at the skin surface, while the steady-state measurement has been specified as the ability of the skin to deliver water to the surface (13,14). Other studies have been done in which the inlet of air has been adjusted to different relative humidities to measure water loss as a function of relative humidity. As the inlet water vapor pressure approaches the outlet water vapor pressure, electrolytic sensors lack the sensitivity to measure the differential water uptake as the air passes through the cell. Other difficulties with this method include a long time to steady state and operator sophistication needed to maintain the system as an operational entity.

Open Cell Gradient Measurement of Transepidermal Water Loss

Several techniques using vapor gradient have been published for evaluation of water loss from the skin surface (15,16). The measurement of water gradient at the skin surface is derived from the equation for one-dimensional diffusion in a medium bounded by two planes, the problem of diffusion from a plane sheet. In the case of diffusion through a plane sheet of membrane thickness L and diffusion coefficient D, where the surfaces are represented by X = 0 at the skin surface and X = L a distance above the skin surface, a constant profile of water vapor concentration is maintained at the steady state. The diffusion Eq. (2) in one dimension reduces to

$$\frac{d^2C}{dx^2} = 0 \tag{3}$$

If we assume that the diffusion coefficient for water is constant in the air above the skin, integration with respect to X yields

$$\frac{-dC}{dx} = \text{constant} \tag{4}$$

This equation shows that the concentration changes linearly from the skin surface to a point a short distance from the skin. Also, the rate of transfer of water vapor is the same across all sections of the membrane and is given by

$$TEWL = \frac{D_{water}(C_2 - C_1)}{l} \tag{5}$$

where l = the distance between the sensors at $x_2 - x_1$.

The gradient of water above the skin surface has been shown to be constant over a range of 1.4 cm (15). Therefore, if the water vapor pressure or water contentration in air is measured at two points above the skin surface, then the water gradient can be calculated from Eq. (5). With a single humidity sensor one can take a measurement close to the skin surface and assume that the ambient water vapor pressure represents the concentration of water at 1.2 cm. The two-sensor device simply measures the water gradient at two distances within 1 cm above the skin surface. The air temperature is recorded, and the water vapor pressure used to calculate flux. In addition, a more accurate measurement can be made by adjusting the diffusion coefficient of water for changes in air temperature. However, changes in the diffusion coefficient over physiological temperature ranges are relatively minor compared to changes in transepidermal water loss when perspiration become a factor at 28–32°C (12).

The gradient systems for measuring transepidermal water loss are fairly simple to use and provide a rapid measure of water loss from the skin surface. The polymeric electrodes used for assessing water vapor pressure have a reasonably rapid response time and do not tend to become saturated. The electronics are somewhat sophisticated; therefore, these sensors do require periodic calibration.

One factor to be considered with the open cell methods is ambient air currents that will disrupt the water gradient close to the skin surface. With a single sensor ambient air currents will make calculations of the water gradient inaccurate. In the case of dual sensor devices, ambient air currents promote rapid fluctuation of the water loss reading, such that obtaining a steady-state value takes an extended period of time. The difficulty with air currents can be accommodated by use of a chimney (17) or a cap (18) over the probe. However, any alterations of the sensor probe also have the potential to alter the water gradient profile above the skin, as discussed later.

APPLICATIONS OF VARIOUS WATER LOSS METHODS

Gravimetrics

Although gravimetric measurements have been used to measure transepidermal water loss from the whole body and from babies in incubators, devices like the dual-sensor gradient measurement are simpler for assessing water loss. The sweat pad method for gravimetric evaluation of perspiration in environmental chambers is the standardized method for measurement of antiperspirant efficacy. The typical protocol employs individuals who will sweat when placed in a chamber at 37°C and who have equilateral sweat in right and left armpits. Individuals use an antipersperant product under one arm for a single use or over a period of days during which they are tested once or repetitively by entering an environmental chamber for a specified period of time at a temperature at which they will perspire profusely. One axilla is used as a control while the other axilla is treated, and the pads are weighed before and after application to the individual axilla. This particular method has proven quite effective in evaluating differences between various antiperspirant products. Even though the method is somewhat laborious in collection, isolation, and weighing of sweat pads, the method does overcome difficulties associated with hair in the armpit that might make other means of measuring water transport more difficult. In addition, frank perspiration tends to overload other means of measuring water loss.

Application of Closed Cell Methods

Closed cell methods have been used to evaluate transepidermal moisture loss in normal individuals and transepidermal water loss from individuals treated with various cosmetic products (19—21). Useful applications of closed cell methods include the ability to monitor an average moisture loss over an extended period of time by applying the cell to the skin and allowing the individual to go about normal activities. Difficulties in this process include assessing small amounts of water in large amounts of hygroscopic material. In addition, closed cell methodology requires subjects who are willing and able to participate in the study. Evaluation of the effects of anticholinergics for reducing insensible perspiration has been demonstrated using closed cell techniques (21). In general, closed cell techniques have been superseded by other methodologies.

Flow-Through Cell

Flow-through cells have been used to evaluate the transepidermal water loss from individuals under many types of stress, including temperature, humidity, and damage by acids, salts, and other chemicals that impair barrier function (23—25). In addition, flow-through open cell techniques have been used to evaluate the correlations between transepidermal water loss and the loss of repellents from the skin surface (9). In these studies, water loss is a reflection of skin permeability or of the water vapor effect on the evaporation of repellent from the skin surface. Flow-through cells have been used in other ways to evaluate cosmetic products (19) and pathological skin conditions (26—29). In general, these applications of flow-through cells have been superseded by the gradient methods, which are more commonly used today.

Since the flow-through methods are continuous measuring systems and tend to measure the delivery rate of water to and through stratum corneum membrane, flow-through methods are practical for evaluation of transepidermal water loss in vitro. One application is measuring moisture transport across skin or an artificial membrane in in vitro permeability cells. Screening of penetration enhancers and of cosmetic products can be accomplished in in vitro cells (30–32).

One unique application of flow-through methods is measurement of the water content at the skin surface. In studies by Cunico et al., a Meeco Electrolytic Analyzer was used to evaluate water loss from the skin surface (14). In the process, the initial water flux from the skin surface prior to reaching steady state was integrated to provide a measure of the water available at the skin surface at the beginning of the experiment. This assessment was accomplished by using the steady-state water loss obtained after 20–30 min as a baseline for the large initial loss of water from the skin surface. This particular measurement is appropriate for evaluation of skin surface moisturization by cosmetic products or topical drug vehicles. In addition, this methodology can be used to evaluate the water retention properties of atopic skin or skin with an incompletely developed stratum corneum, as in the case of psoriasis. The flow cells have also been used in a wipe test to evaluate water at the skin surface underneath occlusive topical products (19).

Application of Gradient Method

Currently the evaporimeter exhibits the widest range of potential uses for measuring water loss. Wound healing is probably the most obvious application (33,34) in which the evaporimeter is used to measure water loss from open wounds. The initial epithelial covering of the wound is accompanied by reductions in water loss and the return of the stratum corneum to a normal barrier function. A variant on the wound healing measurement is that of measuring damage to chapped skin and of monitoring the healing of skin with applications of lotion, as shown in Figure 1. In this study chapped skin exhibited water loss similar to that of partially wounded skin in which the barrier function was severely disrupted. Application of a lotion to the severely chapped skin over a period of days restores the barrier function and provides water loss or barrier function similar to that of normal skin. At the same time that the barrier function is normalized, skin condition is also normalized.

Evaporimetry has been used to measure the barrier function in ichthyotic skin, psoriatic skin, atopic skin, and dry skin (35–38). These various skin conditions do affect barrier function and the serverity of disease is in part measured by the loss of barrier function. The advantage of the evaporimeter in these patient

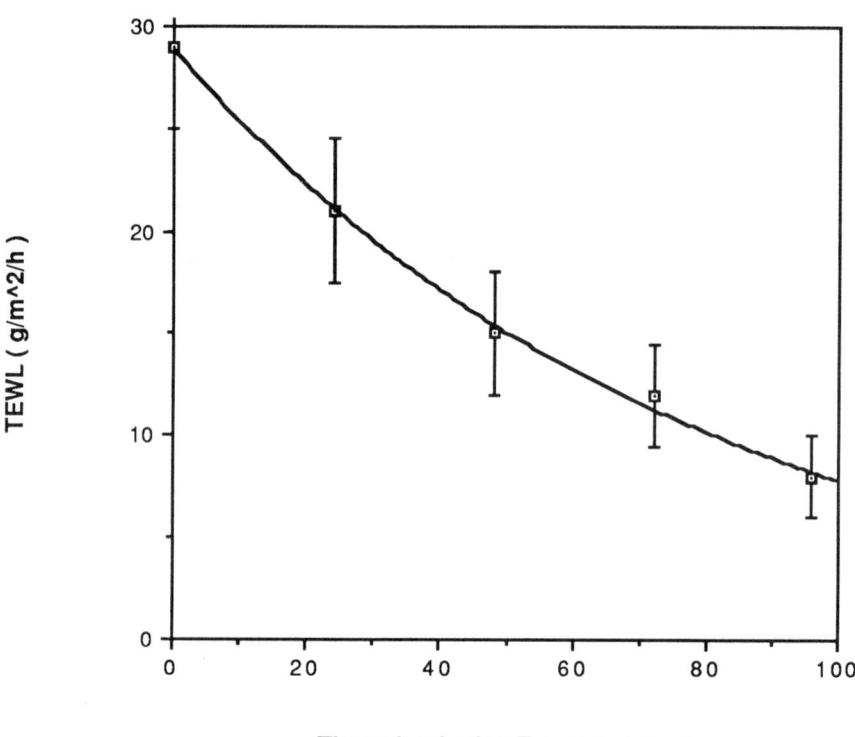

Figure 1 Healing of dry skin with applications of lotion.

measurements is that the TEWL can be measured simply by placing the evaporimeter probe on the skin surface.

A more recent application of evaporimetry has been the monitoring of barrier function as an indicator of penetration enhancement (39). Vinson et al. used closed cell techniques to evaluate the water loss in a series of in vitro studies on skin permeability (31). The evaporimeter has made these measurements more readily available for evaluating the barrier function of stratum corneum prior to beginning in vitro permeability studies. In this way, the investigator knows the competency of the barrier function of the stratum corneum prior to carrying out in vitro permeability experiments. The evaporimeter has also been used to correlate water loss measurements with the effect of penetration enhancers for improving skin permeability.

USE OF THE EVAPORIMETER TO MEASURE TRANSEPIDERMAL WATER LOSS

As discussed previously, steady-state water evaporation from the skin surface creates a gradient concentration from the skin surface to the ambient relative humidity or water vapor pressure at a distance approximately 1–2 cm above the skin surface, depending on ambient air currents (15,16). Within 1 cm of the surface where the gradient is linear, transepidermal water loss can be determined from Eq. (5). The water vapor pressure is measured at two points above the skin surface in the linear gradient to calculate the transepidermal water loss.

One device for calculating transepidermal water loss from this water gradient is the ServoMed Evaporimeter. The evaporimeter probe is a small cylinder with two capacitive thin film transducers approximately 1/2 cm apart for measuring the relative humidity above the skin (Fig. 2). Adjacent to each transducer is a thermistor for measuring air temperature. The signal from these sensors is translated to a partial vapor pressure of water, corrected for temperature; then, using a preset algorithm, the partial pressure gradient and evaporation rate are recorded. Specifications for the instrument call for an accuracy in transepidermal water loss of ±15%, if the probe is left on the skin for 5 min in a temperature range of 18–35°C. With some alteration to the standard procedure discussed below, reduction in the time to a steady-state

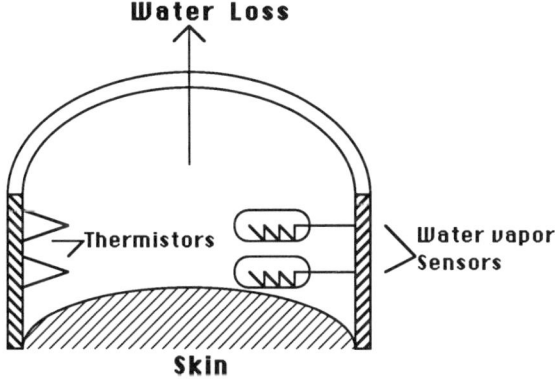

Figure 2 ServoMed Evaporimeter.

reading and improvement in the reproducibility is observed. Equilibration time is necessary both for response of the relative humidity sensors and for attainment of temperature equilibrium around the probes (44). In addition, air currents disrupt the linear gradient above the skin surface, causing wide fluctuations in water loss measurement until steady state is obtained.

Air Currents

Using gradient methods for evaporimetry is made more difficult by normal airflow that disrupts the gradients above the skin, making steady state difficult to attain. This is especially true in environmentally controlled chambers where temperatures and relative humidity are controlled via air-conditioning systems that circulate air rapidly. With the ServoMed Evaporimeter, up to 5 min is required for stabilization of temperature and ambient air currents to allow a quasi-steady-state reading of transepidermal water loss. One still observes moderately rapid fluctuations in the water loss digital readout from the evaporimeter. In environmental chambers the fluctuation is even more rapid. Two approaches have been designed to accommodate the environmental air currents: (1) placement of a cap with hygroscopic salt above the evaporimeter sensor cylinder (18) and (2) placement of a chimney extending the air current above the evaporimeter cylinder (17). A cap with a hygroscopic salt above the cylinder provides a closed environment for measuring the water gradient above the skin with an infinite sink for the moisture being given off into the atmosphere. The method eliminates air currents completely and establishes a reproducible constant gradient from the skin surface to a 0% relative humidity controlled environment. The compromise using this procedure is that the ambient environment is always 0% relative humidity. In many cases, there is an interest in studying water loss from the skin surface into ambient air of various relative humidities. To accomplish measurement of transepidermal water loss ambient at relative humidity, the air currents around the sensor must be controlled.

The second means of controlling air currents around the sensor is to provide a chimney extending the column of static air above the sensors by 1–2 cm. Initial suggestions for eliminating air currents at the sensor include placing a plastic syringe barrel on the probe, forming a continuous cylinder extending 3–4 cm above the probe. This was followed by a more complete design using a Teflon chimney that was adapted to slide onto the probe, extending the column of air above the sensors by approximately

Table 1 Effect of Temperature and Chimney on Time to Obtain Steady-State Water Loss

Conditions for sensor	TEWL ($g/m^2/hr$) (mean ± std. dev.)	Time to steady state
Room temperature 21°C, no chimney	5.3 ± 2.6	>5 min Rapid fluctuation
Skin temperature 29.5°C, no chimney	5.4 ± 2.0	>5 min Rapid fluctuation
Room temperature 21°C with chimney	4.7 ± 2.5	90 sec Stable
Skin temperature 29.5°C, with chimney	6.2 ± 2.2	60 sec Stable

1.5 cm. The chimney reduces the time to steady-state equilibrium from 5 min to approximately 90 sec (18; Table 1). However, the extended column of air creates a slight decrease in the water gradient that results in a slight, but reproducible, decrease in the transepidermal water loss determination (Table 2). Since the slight depression is reproducible and constant in the range of the transepidermal water loss around normal skin ($0-20$ g/m^2), the chimney makes a highly acceptable mechanism for most applications of the evaporimeter. In evaluation of wound healing where the range of TEWL extends to 100–150 $g/m^2/hr$, the suppression of TEWL may be significant; however, in studies of wound healing the formation of a partial barrier to transepidermal water loss during the healing process causes quite dramatic drops in the observed transepidermal water loss and brings the measurement rapidly into a range in which the suppression is acceptable.

Temperature Effects

Temperature effects on transepidermal water loss measurements are a function of the subject on which the measurement is being made and the instrumentation. Ambient temperature has been demonstrated to affect transepidermal water loss measurements in a reproducible fashion (12,40–43). Several authors have

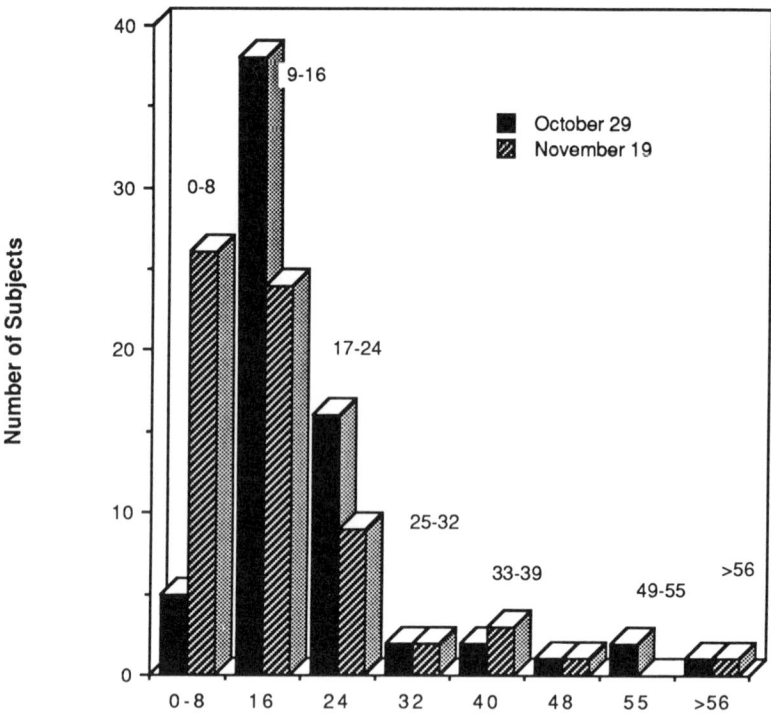

Figure 3 Wide variation of water loss in a large number of subjects in the uncontrolled environment of a clinical study.

worked out correction factors for transepidermal water loss and skin temperature (43). Transepidermal water loss decreases as skin temperature decreases, especially under circumstances where ambient temperatures fall significantly below 20°C. At ambient temperatures between 25 and 32°C different subjects have differing responses in increase of transepidermal water loss. The temperature effects are no longer predictable because subjects start to perspire at varying temperatures (Fig. 3). At approximately 32°C most subjects perspire, and resulting transepidermal water loss is high. In order to minimize variation due to temperature, the investigator should record skin temperature as a routine part of transepidermal water loss measurements. In order to accommodate variations in transepidermal water loss due to changes in ambient temperature and skin temperature, the room in which measurements are taken should be kept at a constant 20–23°C. In addition, to accommodate variation in skin temperature between individuals, subjects should be equilibrated for a period of time prior to the study (approximately 20 min), and control measurement sites should be included in the experimental design for each subject. The controls and equilibration period will allow subjects, who have rushed to the test center for their appointment, time to cool down and relax before transepidermal water loss measurements are taken.

The second temperature factor in transepidermal water loss is the effect of temperature on the probe. A probe placed on the skin at room temperature of 20–23°C will require a period of time to equilibrate to approximately skin temperature before steady state is obtained. This factor is especially important in gradient measurements in which the water concentration gradient above skin is temperature dependent. If ambient air currents are eliminated from consideration by use of a chimney above the evaporimeter probe, the equilibration time for a steady-state water loss measurement is approximately 60–90 sec (Table 1). If, however, the probe is warmed to 29°C, the time decreases to 30–60 sec (17,43). A simple means for keeping the probe at a constant

Table 2 Effect of Chimney Height on Transepidermal Water Loss

Chimney height (cm)	Water loss ($g/m^2/hr$)
0	7.51
0.5	5.67
1.0	5.41
2.0	5.11
4.0	5.01

temperature is an aluminum block heated to the temperature selected as approximately skin temperature. In situations where it is not possible to equilibrate the probe at a temperature close to skin temperature, the investigator must allow a slightly longer time to reach steady state.

A final aspect of temperature is the external ambient temperature outside the test center. Subjects who come into a test center from outside may at times be exposed to high heat and humidity or sometimes extremely cold temperatures in the hour prior to arrival. The body temperature and skin temperature adapt to these external ambient conditions. Under circumstances in which the external ambient temperature is extremely hot or extremely cold, longer times for subject equilibration prior to making water loss measurement are necessary to allow the skin temperature to come to a reproducible equilibrium. The factor of occlusive clothing or other material worn on the skin in the winter should also be considered because stockings, boots, and heavy wool socks may alter skin surface water content, if appropriate time is not made for equilibration. In addition, in the summer months, sun exposure leads to sunburn, which alters skin temperature and skin water loss.

Relative Humidity Effects

Supply of moisture to the stratum corneum from the dermal layers is sufficient to create a significant water content at the skin surface for normal skin. The gradient from the skin surface to an ambient relative humidity of 20–40% is a steep gradient appropriate for consistent transepidermal water loss measurements. As the relative humidity in the ambient air increases, the gradient or potential for moisture transport across the skin and into the ambient air is decreased; therefore, as expected, the transepidermal water loss decreases. Several investigators have observed a downward trend in transepidermal water loss between 30 and 80% relative humidity using different methods of measuring transepidermal water loss (45–48). At very high relative humidities the flow methods are not sensitive enough to measure the difference between incoming and outflowing air after it crosses the skin surface. Likewise, in the case of the evaporimeter the measurement of water loss becomes more difficult at relative humidities above 90%. Some studies have indicated a drop in transepidermal water loss at low-level humidities (45), while other studies indicate that the transepidermal water loss increases at low humidities (46–48).

For standardized measurements where temperature is controlled, a relative humidity from 20 to 40% is the most desirable. This relative humidity range is one in which subjects will be reasonably comfortable at the low temperatures (20–23°C) preferred for water loss measurements. Because high relative humidity in summer or exceptionally low relative humidity in winter affects TEWL measurements, studies should be designed so that the measurements in extreme environments are part of a separate block, which can be analyzed independently for variation and then pooled with larger sets of data. Although high relative humidities will not adversely affect evaporimeter measurements, it is possible that electrolytic cells relying on conductivity of salt bridges will be saturated by extremely high relative humidity.

Subject Selection and Preparation

Individual variation is critical to the success and reproducibility of water loss studies; consequently, subject selection is important. First, subjects should be selected who are appropriate to the measurement being taken. If skin diseases are of interest, subjects who have been previously diagnosed for those diseases are one phase of the study, while control subjects with normal skin become a second measurement group. Care should be taken in selection of control subjects to be sure that these subjects are not suffering from conditions such as atopic dermatitis in which uninvolved skin may appear normal to the casual observer. In the hospital setting, it is quite common to find subjects with skin conditions that are not apparent clinically.

At the time of entry into the study, subject age, sex, physical condition, and general skin condition should be noted. If the subject is taking any medications or has any unusual dietary or exercise habits, those should be recorded on entry into the study. The latter factors should be considered in the event that unusual data are observed during the study. For instance, in the case of an individual who is very active in outdoor athletics, in a study conducted in summer, a slight exposure to sun or strenuous exercise on the day of a study could increase TEWL. Internal controls for each subject alleviate the situation to a certain extent.

Even under the best circumstances, there is day-to-day variation within a given subject. These changes are affected by physiological changes within the individual as well as changes in the external environment. Within-subject day-to-day variation is controlled as best as possible using multiple sites on a single individual. In the case of skin diseases, a control site of uninvolved skin close to the involved site is sometimes an appropriate control. Studies have shown that there is a successive decrease in transepidermal water loss moving down the volar forearm. The change is not large; however, if small differences are being studied, the change is significant. In addition, an individual may have slightly higher transepidermal water loss on the right or left forearm, eliminating the concept of strict bilateral symmetry. Right or left dominance determines which forearm has the higher transepidermal water loss. The right-left difference may be accentuated if the individual has participated in a physical endeavor requiring use of his dominant hand just prior to the start of the study.

Subject excitability or emotional state is important in making transepidermal water loss measurements. In circumstances where a patient is being evaluated, he should be allowed to ask any questions and be placed at ease when the measurements are being taken. If caution is not taken to place the patient or test subject at ease, emotional stress may lead to nervous perspiration, elevating transepidermal water loss measurements. In clinical tests of topical products or surveys, subjects should be allowed to recline at rest in an isolated or partitioned room. An example is shown in Figure 3 of a study conducted on hand lotion products in which the subjects were measured in an open room under sometimes chaotic conditions. Note that the range of transepidermal water loss values observed is quite broad. Although a larger number of subjects permits demonstration of statistically significant differences in transepidermal water loss between treatments, one observes a number of water loss values above 30 $g/m^2/hr$. In this particular study, higher values result from inadequate equilibration time, the wearing of boots by some subjects, a warm desert environment, and the uncontrolled emotional stress of the test center environment. In the case of one set of measurements made just before a holiday, two subjects exhibited high emotional stress as indicated by a sudden elevation in water loss when the subject of holidays arose. Therefore, the operator must avoid any significant conversation with the subject outside of casual polite comments to place the subject at ease. In large institutions the management factor is important in reducing subject stress. If managers enter the study room during a water loss measurement, most subjects experience some degree of emotional stress, and TEWL will require 20 min to come back to equilibrium.

Test Facility

Any room that can be maintained at a reasonably constant temperature below 25°C and wihin a reasonable relative humidity range between 20 and 50% can be used to take water loss measurements. In clinical or test center situations where control of ambient conditions is not possible, care should be taken to record the exact ambient conditions at the time measurements are taken. In general, a separate room should be available for water loss measurements. If possible, an area should be set aside where the subject can equilibrate for a period of time prior to measurement of transepidermal water loss. During this time the subject should not be engaging in highly animated conversation with large numbers of people, since these conversations could lead to emotional stress. Where possible, the test room should be reasonably soundproofed. If there are windows in the test facility, the subject should be seated so as not to be looking out the window. Additionally, he should not be able to see the water loss readings on the digital display.

Environmental rooms can be constructed using a room air conditioner, coupled with a dehumidifer system. The operation of an air conditioner and/or dehumidifier also serves to provide background noise that eliminates external stimuli for the subject. Regardless of the conditions that can be maintained in a facility, the most important factor is the ability to reproduce constant conditions for taking transepidermal water loss measurements. Even in situations where conditions cannot be controlled, the numbers of subjects can be increased to accommodate the increased statistical variation.

Study Design

One of the most important aspects of study design is the experimental objective. Once an objective is defined, the rest of the study and the variables associated with transepidermal water loss can be controlled or confounded depending on the objective of the experiment. Because of the large number of variables in transepidermal water loss measurements, lack of an adequate definition of the experimental objective may cause a large number of subjects to be used in a poor experimental design. In addition to the objective, factors in study design include (1) sites of applications, (2) selection of subjects, (3) number of subjects, (4) statistical design of the experiment, (5) time for equilibration, (6) inter- and intrasubject variability, (7) control of environmental condition, and (8) ability to control the subject during the study, i.e., time for subject equilibration, time for measurements, and availability of the subject to remain in a controlled environment between measurements. Where possible, multiple sites for water loss measurements should be used on each subject to provide an intrasubject control or even bilateral controls on the same subject. Where studies of skin diseases are the objective, noninvolved skin in an area adjacent to the involved area is often an appropriate control and a possible normal control group may be used. However, the normal controls should be truly normal. Selections of normals from hospital patient populations may provide large numbers of individuals with atopic dermatitis (35), uninvolved psoriatic skin (42), or diabetes, all of which may affect transepidermal water loss measurements. As many causes as possible of variation in water loss measurements should be controlled in the experimental design or, alternatively, larger number of subjects should be included to enhance the possibilities of seeing significant differences.

In studies of topically applied products, the drying time for creams and lotions should be considered in the selection time following initial application for measuring transepidermal water loss. The results of applying a lotion to the skin and monitoring transepidermal water loss can be seen in Table 3 in which product A and product B are moderately occlusive moisturizing lotions that initially increase transepidermal water loss, followed by a reduction in water loss compared to time 0. The control values do not change significantly between time 0 and 1 hr. At 45 min product B still has not exhibited a significant difference from time 0 in transepidermal water loss; however, in a second study in which subjects were allowed 60 min drying time, observed reductions in TEWL were consistent. The number of subjects selected in studies must be determined by the ability to control variables and by consideration of the number of days in which the study will be carried out. In a similar study petrolatum was used as a positive highly occlusive control (Table 3). Using both an untreated and positive control, one can gauge the degree of occlusivity of a product. Another method for evaluating these types of products is to wipe the product from the skin surface after a specified time to determine the increase of moisture loss due to occlusivity, as reported by Rietschel (19).

As mentioned previously, in the case of dry, chapped skin, the degree of damage is similar to a wound, with water loss values on

Table 3 Evaluation of Occlusive Effects with Transepidermal Water Loss

Treatment	TEWL ($g/m^2/hr$) (mean ± std. dev.)	
	Control time 0	60 min
Product A	7.2 ± 2.8	6.4 ± 1.9[a]
Product B	7.3 ± 2.0	6.2 ± 1.6[a]
Untreated control	6.4 ± 1.5	6.2 ± 1.3
Petrolatum control	6.5 ± 1.9	3.5 ± 0.91[a]

[a]Significant difference ($p < 0.05$) by paired comparison with control at time 0 before application, $n = 12$.

the order of 30 $g/m^2/hr$ per hour (Fig. 1). With application of an effective hand and body lotion over a period of days, the barrier function is restored and water loss decreases to nearly normal values. Experimental design in these studies is simplified by the fact that the reduction of water loss is very significant, 20–30 $g/m^2/hr$, relative to the accuracy of the instrument and individual subject variability. If the skin is not very dry, the factor of carrying the study out over several days makes detection of significant differences in TEWL more difficult because of day-to-day subject variability and individual differences in resolution of dry skin by treatment products.

Evaporimeter Instrumentation

The ServoMed Evaporimeter as previously described is available with one or two probes, each with calibrated electronics to provide an accurate measurement of relative humidity or water transpiration for given sensors. The sensors should be calibrated on a regular basis with saturated salt solutions. Lithium chloride, potassium sulfate, and magnesium nitrate provide standard relative humidities of 15%, 93%, and 63%, respectively. The saturated salt solutions are placed in a sealed flask and allowed to come to equilibrium at a constant temperature for at least 24 hr. The sensor probe is inserted quickly into the top of the flask and allowed to come to equilibrium, approximately 1–2 hr. If the flask containing the salt solution remains open for more than a few moments before the probe is inserted, a longer time should be allowed for the equilibrium calibration. The operator should look for trends in water loss values either higher or lower. When water loss values are recognized to be erratic, the probe must be returned to ServoMed for recalibration of the electronics with respect to the sensors. Many investigators obtain ServoMed systems with dual probes, each with its own solid-state electronics; thus, one probe is sent back for calibration while the second probe is used in continuing studies. The probe should be maintained on a rack or resting in a temperature-controlled block to avoid damage. The instrumentation should be positioned such that the investigator can observe steady-state water loss out of the line of sight of the subject being tested. The probe is placed firmly on the subject's skin to prevent motion and to avoid excessive presure, which has been observed to affect the water loss measurement. Some operators use a slight mass applied on the upper side of the probe to provide a constant application force. When measurements are being made on vertical surfaces of the skin, care should be taken not to press too hard or to allow the probe to fall away from the skin surface. Motion of the probe during the measurement allows air currents to disrupt the water gradient and may provide inaccurate water loss measurements. Between uses, the evaporimeter should be stored in an area not exposed to solvent vapors, dust, and dirt. Many laboratories are clean enough that storage of the probe in an open configuration is not a problem.

SUMMARY

Although the focus on measurement of transepidermal water loss has been on use of gradient methods with an evaporimeter, each method of measuring transepidermal water loss has utility in specific circumstances, for example, use of flow methods in measuring skin surface water. The main point to understand in measuring transepidermal water loss is to know the instrument and know the method. Knowledge of specific instrumentation and the limits of the instrumentation will provide the investigator with much greater accuracy in measurements. Following the knowledge of instrumentation, the most important aspect of conducting any study is the definition of objective and the understanding of how treatments or toiletry products will affect the time course of water loss measurements. Finally, a protocol designed to accommodate anticipated variables in the measuring sequence is extremely important, as are the statistics used to analyze results. Understanding the variation associated with transepidermal water loss and the subject/water loss interaction will provide an excellent basis for using water loss measurements for evaluation of skin diseases, skin barrier function, and the effects of topical drugs and toiletries applied to the skin surface.

REFERENCES

1. Reay, D. A., and Thiele, F. A. J. Heat pipe theory applied to a biological system: Quantification of the role of the resting eccrine sweat gland in thermoregulation. *J. Theoret. Biol.* 64:789–803 (1977).
2. Crank, J. Diffusion in a plane sheet. In: *Mathematics of Diffusion*, Oxford/Clarendon Press, London, pp. 42–44 (1975).
3. Wu, M-S., Yee, D. J., and Sullivan, M. E. Effects of a skin moisturizer on the water distribution in human stratum corneum. *J. Invest. Dermatol.* 82:446–448 (1983).
4. Blank, I. H., Moloney, J., III, Emslie, A. G., Simon, I., and Apt, C. The diffusion of water across the stratum corneum as a function of its water content. *J. Invest. Dermatol.* 82:188–194 (1984).
5. Powers, D. H., and Fox, C. The effect of cosmetic emulsions on the stratum corneum. *J. Soc. Cosmet. Chem.* 9:109–116 (1959).
6. Monash, S., and Blank, H. Location and reformation of the epithelial barrier to water vapor. *Arch. Dermatol.* 78:710 (1958).
7. Berube, G. R., Messinger, M., and Berdick, M. Measurement in vivo of transepidermal water loss. *J. Soc. Cosmet. Chem.* 22:361–366 (1971).
8. Quattrone, A. J., and Laden, K. Physical techniques for assessing skin moisturization. *J. Soc. Cosmet. Chem.* 27:607–623 (1976).
9. Rietschel, R. L., and Spencer, T. S. Correlation between mosquito repellent protection time and insensible water loss from skin. *J. Invest. Dermatol.* 65:385–387 (1975).

10. Bettley, F. R., and Grice, K. A. A method for measuring the transepidermal water loss and a means of inactivating sweat glands, *Brit. J. Dermatol.* 77:627–638.

11. Spruit, D. Measurement of the water vapor loss from human skin by a thermal conductivity cell. *J. Appl. Physiol.* 23(6): 994–997 (1967).

12. Thiele, F. A. J., and Malten, K. E. Insensible water loss. Intersubject variation related to skin temperature, forearm circumference and sweat gland activity. *Trans. St. Johns Hosp. Dermatol. Soc.* 58:199 (1972).

13. Lamke, L. O. An instrument for estimating evaporation from small skin surfaces. *Scand. J. Plast. Reconstr. Surg.* 4:1–7 (1970).

14. Cunico, R. L., Maibach, H. I., Kahn, H., and Bloom, E. Skin barrier properties in the newborn: Transepidermal water loss and carbon dioxide emission rates. *Biol. Neonate* 37:180–185 (1977).

15. Nilsson, G. E. Measurement of water exchange through skin. *Med. Biol. Eng. Comput.* 15:209–218 (1977).

16. Oberg, P. A., Hammerlund, K., Nilsson, G. E., Nilsson, L., and Sedin, G. Measurement of water transport through the skin. *Uppsala J. Med. Sci.* 86:23–26 (1981).

17. Seitz, J. C., and Spencer, T. S. Use of capacitive evaporimetry to measure effect of topical ingredients on transepidermal water loss. *J. Invest. Dermatol.* 78:351 (1982).

18. Bowman, W. D. In vitro and in vivo assessment of human stratum corneum barrier function by transepidermal water flux. *Proceedings of the 4th International Symposium of Bioengineering and Skin.* Besancon, France, MTP Press, Lancaster, p. 38 (1983).

19. Rietschel, R. L. A method to evaluate skin moisturizers in vivo. *J. Invest. Dermatol.* 70:152–155 (1978).

20. Leveque, J. L., Garson, J. C., and de Rigal, J. Transepidermal water loss from dry normal skin. *J. Soc. Cosmet. Chem.* 30:333–343 (1979).

21. Pinsi, E. A. Evaporation from human skin with sweat glands inactivated. *Am. J. Physiol.* 137:492 (1942).

22. Hattingly, J. The correlation between transepidermal water loss and the thickness of epidermal components. *Comp. Biochem. Physiol.* 43A:719–722 (1972).
23. Malten, K. E., and Spruit, D. Injury to the skin by alkalai and its regeneration. *Dermatologica* 132:124–130 (1966).
24. Valk, P. G. M. v.d., Nater, J. P., and Bleumink, E. Skin vapor loss as a method for measuring the influence of soaps and detergents on human skin. *Dermatosen* 31:58–60 (1983).
25. Hoffmann, H., and Maibach, H. I. Transepidermal water loss in adhesive tape induced dermatitis. *Contact Dermatitis* 2:171–177 (1976).
26. Rajka, G., and Thune, P. The relationship between the course of psoriasis and transepidermal water loss, photoelectric plethysmography and reflex photometry. *Br. J. Dermatol.* 94:253–261 (1976).
27. Grice, K., Sattar, H., Baker, H., and Sharratt, M. The relationship of transepidermal water loss to skin temperature in psoriasis and eczema. *J. Invest. Dermatol.* 64:313–315 (1975).
28. Abe, T., Ohkido, M., and Yamamoto, K. Studies on skin surface barrier functions. *J. Dermatol.* 5:223–229 (1978).
29. Rajka, G. Transepidermal water loss on the hands in atopic dermatitis. *Arch. Dermatol. Forsch.* 251:111–115 (1974).
30. Onken, H. D., and Moyer, C. A. The water barrier in human epidermis. *Arch. Dermatol.* 87:584–590 (1963).
31. Vinson, L., Koehler, W. R., Masurat, T., and Singer, E. A. Basic studies in percutaneous absorption. Report No. 6, Contract Number DA18-108-CML-6573.AD 434 459. U.S. Clearinghouse for Federal Scientific Information, U.S. Department of Commerce, Washington, DC (December 1963).
32. Berube, G. R., and Tranner, F. Transepidermal Moisture Loss. III. An in vitro approach. *J. Soc. Cosmet. Chem.* 30:181–190 (1979).
33. Salisbury, R. E., Carnes, R. W., and Enterline, D. Biological dressings and evaporative water loss from burn wounds. *Ann. Plastic Surg.* 5:270–272 (1980).

34. Lamke, L-O., Nilsson, G. E., and Reithner, H. L. The evaporative water loss from burns and the water-vapour permeability of grafts and artificial membranes used in the treatment of burns. *Burns* 3:159–165 (1977).

35. Werner, Y., and Lindberg, M. Transepidermal water loss in dry and clinically normal skin in patients with atopic dermatitis. *Acta Dermatol. Venereol. (Stockh.)* 65:102–105 (1985).

36. Werner, Y., and Lindberg, M. Transepidermal water loss in dry and clinically normal skin in patients with atopic dermatitis. *Acta Dermatol. Venereol. (Stockh.)* 65:102–105 (1985).

37. Leveque, J. L., Grove, G., et al. Biophysical characterization of dry facial skin. *J. Soc. Cosmet. Chem.* 82:171–177 (1987).

38. Van Neste, D., Masmoudi, B., Leroy, B., Mahmoud, G., and LaChapelle, J. M. Regression patterns of transepidermal water loss and of cutaneous blood flow values in sodium lauryl sulfate induced irritation: A model of rough dermatitic skin. *J. Bioeng. Skin* 2:103–118 (1986).

39. Dupuis, D., Rougier, A., Lotte, C., et al. In vivo relationship between percutaneous absorption and transepidermal water loss according to anatomic site in man. *J. Soc. Cosmet. Chem.* 37:351–357 (1986).

40. Lamke, L-O., Nilsson, G. E., and Reithner, H. L. Insensible perspiration from the skin under standardized environmental conditions. *Scand. J. Clin. Lab. Invest.* 37:325–331 (1977).

41. Lamke, L-O., and Wedin, B. Water evaporation from normal skin under different environmental conditions. *Acta Dermatol. Venereol. (Stockh.)* 51:111–119 (1971).

42. Grice, K., Sattar, H., Baker, H., and Sharratt, M. The relationship of transepidermal water loss to skin temperature in psoriasis and eczema. *J. Invest. Dermatol.* 64:313–315 (1975).

43. Mathias, C. G., Wilson, D. M., and Maibach, H. I. Transepidermal water loss as a function of skin surface temperature. *J. Invest. Dermatol.* 77:219–220 (1981).

44. Blichmann, C. W., and Serup, J. Reproducibility and variability of transepidermal water loss measurement. *Acta Dermatol. Venereol. (Stockh.)* 67:206–210 (1987).

45. Grice, K., Sattar, H., and Baker, H. The effect of ambient humidity on transepidermal water loss. *J. Invest. Dermatol.* 58:343–346 (1972).

46. Goodman, A. B., and Wolf, A. V. Insensible water loss from human skin as a function of ambient vapor concentration. *J. Appl. Physiol.* 26:203–206 (1969).

47. Hammarlund, K., Nilsson, G. E., Oberg, P. A., and Sedin, G. Transepidermal water loss in newborn infants. I. Relation to ambient humidity and site of measurement and estimation of total transepidermal water loss. *Acta Pediatr. Scand.* 66:553–562 (1977).

48. Sauer, P. J., Dane, H. J., and Visser, K. A. Influence of variations in the ambient humidity on insensible water loss and thermoneutral environment of low birth weight infants. *Acta Pediatr. Scand.* 73:615–619 (1984).

12
Assessing Hair Growth in Male Pattern Baldness

ROBERT L. RIETSCHEL *Ochsner Clinic, New Orleans, Louisiana*

INTRODUCTION

The development and testing of a topical minoxidil solution for the treatment of male pattern baldness (androgenetic alopecia) have focused attention on quantitating the degree of success that can be obtained by this and other drugs (1-9). Since hair and its care and styling are part of our personal appearance, a high level of subjectivity in the grading of hair growth is likely. This was borne out in the 27-center trials of Rogaine in the United States (1-9). When patients and investigators were asked to rate hair growth as none, mild, milderate, or dense, approximately 50% of patients rated their hair growth as moderate after 12 months, while the investigators rated hair growth moderate in only 40% of the same individuals.

Direct hair counts taken from a defined area offer an objective measure of response regarding quantity but not quality. The area selected for counting in initial studies focused on the center of the balding vertex of the scalp. As treatment progressed, it was noted that the central vertex was often slower to respond than the peripheral vertex regions. The response of the vertex did not always predict response of the anterior hairline. Therefore, the first decision to be made prior to performing hair counts is whether the study addresses the anterior hairline or occipital (vertex) bald spot.

The second decision is how the same area is to be identified throughout the study period. In early Rogaine trails, a template cut from a thin sheet of clear plastic was used to locate by triangulation a site on the center of the balding area. The anatomical landmarks used to align the template at each visit were the tip of the nose and the tops of the ears. With these three points of reference, the same spot could be identified for counting with reasonable reproducibility. However, with this method, small differences in orientation will make the data more variable than necessary. The preferred method is direct tattooing of the skin for exact identification of the treatment site to assure that placement of the template is correct from visit to visit. If a circular site is desired, a single dot will suffice to mark the center of the study site. From the central dot a circle can be constructed with a compass. If a square configuration is desired, two dots are necessary to assure proper alignment (preferably at opposite diagonal corners). The circular method can also be constructed by a semicircular template rather than with a compass. The template is shown in Figure 1. A central notch is made to align with the central dot. Making the semicircular template greater than 180° facilitates completion of the circle because the template must be rotated 180° to draw a complete circle. A pen with a roller tip or firm fiber or felt tip can be used with either a semicircular or square template to draw an outline on the skin.

The permanent marks can be made by placing a drop of India ink at the site selected and piercing the drop with a standard 21-gauge needle used for drawing blood. The skin need be pierced only to a depth of 2 mm. Gauze should be pressed firmly against the skin after needle puncture because bleeding is common. Then the excess ink is wiped away. The dot should be about the size of a standard typewriter period. It is permanent but can be removed with dermabrasion. Since the dots made in this way would be apparent on totally bald skin, a template using anatomical landmarks still has a place in the study of the anterior hairline.

Now that an identifiable site for hair counting is present, hairs may be counted manually; however, a photographic record is desirable. The photograph can be enlarged or projected to facilitate

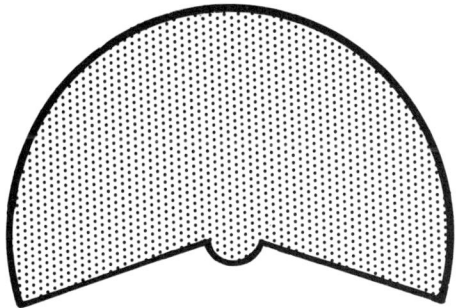

Figure 1 Semicircular marking template.

counting. There are inexpensive ways to obtain good closeup photographs. Extension tubes (one or several) can be placed between the standard 50-mm lens and the body of a 35-mm SLR camera to provide excellent enlargement, and gauze or tissue can be placed over a side- or top-mounted flash to act as a diffuser. A test roll or two of the selected film will provide the proper f-stop and shutter speed. The focal length should be kept constant to provide comparable photographs from session to session. More elaborate camera systems can be designed to suit individual tastes.

Armed with the objective measure of hair counts, we are still limited in our discussion of results. The density of hair growth that constitutes the full normal amount of hair for any given man is needed to give perspective to hair counts. If hair growth within a target area increases from 50 hairs per cm^2 to 200 hairs per cm^2, is the growth good, bad, or indifferent, and by whose taste? The solution is to determine, prior to the study, normal hair density in the same size area from scalp that does not manifest the trait of male pattern baldness, i.e., the area of the posterior scalp below the level of the tops of the ears. This area is used for hair transplantation donor sites and seems to represent the density that, if achieved, would entirely overcome the manifestation of male pattern balding. This area is influenced by aging in all men and does thin from puberty to senescence.

Assume a 45-year-old man has moderate (Hamilton class V) balding and no hair at all in the target area on the vertex of his scalp selected for study. The density of hair at a site 4 cm behind the left postauricular sulcus is determined to be 100 cm^2.

After a defined period of treatment, the hair count on the scalp vertex is 50 hairs per cm^2. We can say that the treatment produced a 50% recovery of the hair density that would be expected in this patient at his age. If the treatment produced 100 hairs per cm^2, then male pattern baldness would have been overcome in this individual. If hair growth in the treatment area exceeds the density of hair growth at the control site, we can presume that the treatment is effective against hair loss from male pattern baldness as well as from aging.

Even with hair counts evaluated and baseline hair density determined as noted above, the quality of the hair should be assessed. If all hair is vellus (baby fine, short, unpigmented, peach fuzz), then the benefit will be distinctly different from the benefit resulting from growth of coarse, pigmented terminal hair. The two template-counted areas can again be assessed for the number of vellus and terminal hairs present. A ratio of terminal hairs divided by vellus hairs can be used to assess quality of growth, with larger numbers equated with better quality.

As hair growth makes accurate photography and counting difficult because the hairs overlap and obscure each other, it is necessary to cut the hair at the study sites to obtain accurate data. With a round or square plastic template placed over the marked sites, the hair in the area is pulled through the opening with a crochet hook and cut to a length of 1–2 mm. With 10× magnification the hair can be accurately counted and classified. Hair normally grows about 10–14 mm every 4 weeks, and this creates a suitable time interval between visits during the study.

The study design should include a placebo treatment because placebo responses, particularly of vellus hair growth, have been commonly reported (1,6). The length of placebo control should equal or exceed by at least 2 months the length of time anticipated to achieve significant hair growth with the experimental agent. Studies of currently available agents should continue for at least 8–12 months. More prolonged treatment and study are valuable, because a lifetime of use would be anticipated and long-term benefit can only be identified by long-term study.

REFERENCES

1. DeVillez, R. L. J. Am. Acad. Dermatol. 16:669 (1987).
2. Shupack, J. L., Kassimer, J. J., Thiramoorthy, T., Reed, M. L., and Jondreau, L. J. Am. Acad. Dermatol. 16:673 (1987).

3. Rietschel, R. L., and Duncan, S. H. *J. Am. Acad. Dermatol.* 16:677 (1987).
4. Kassimir, J. J. *J. Am. Acad. Dermatol.* 16:685 (1987).
5. Olsen, E. A., DeLong, E. R., and Weiner, M. S. *J. Am. Acad. Dermatol.* 16:688 (1987).
6. Savin, R. C. *J. Am. Acad. Dermatol.* 16:696 (1987).
7. Roberts, J. L. *J. Am. Acad. Dermatol.* 16:705 (1987).
8. Katz, H. I., Hien, N. T., Prawer, S. E., and Goldman, S. J. *J. Am. Acad. Dermatol.* 16:711 (1987).
9. Kreindler, T. G. *J. Am. Acad. Dermatol.* 16:718 (1987).

Index

Abrasion,
  animal, 25-26
Acne, 185-189
  free fatty acids, 187-188
  global scoring, 185-187
  impression techniques, 188
  lesion,
    counting, 185-187
    definitions, 186
  lipid composition, 187-188
  microflora, 155-157, 162, 188
  sebum excretion rates, 187

Bacterial sampling, 143-164
  air sampling, 161
  biopsy techniques, 146, 156-160
  in diseases, 160-161
  impression techniques, 145-150
  recommendations, 161-164

[Bacterial sampling]
  surface techniques, 147, 155-156
  swabbing techniques, 146, 153-155
  washing techniques, 146, 150-153
Blood flow, 107-116
  clinical studies, 112-113
  experimental studies, 113-114
  Laser-Doppler velocimetry, 108-110
  photopulse plethysmography, 110-112
Buehler test, 51, 53

Contact dermatitis,
  allergic (see Sensitization assays)
  irritant (see Irritant testing)

225

Dossou and Secard test, 51
Draize test, 20-22, 49

Ear flank test, 51, 53
Epilation, 25
Evaporimeter, 30-31, 199-203, 211-212
Experimentation, human, 1-16
　basic principles, 4-6
　forms required, 15-16
　Helsinki declaration, 3-4
　historical perspective, 2-3
　informed consent, 5, 10, 12-13
　minimal risk, 9
　non-therapeutic, 6
　phase I, II, III trials, 8
　protection of subjects, 7-9

Federal Hazardous Substance Act, 20-22
Freund's complete adjuvant test, 51, 52

Grotthus-Draper law, 94
Guinea pig maximization test, 51-55, 66

Hair, 219-222
　control sites, 221-222
　counting, 219-220
　photographic documentation, 220-221
　scalp site identification,
　　tattoo method, 220
　　template method, 220

[Hair]
　types, 222
Helsinki Declaration, 3-4

Immersion test, 32
Impedance measurement, 121-139
　electrodes, 131-132
　experimental conditions, 132-133
　instrumentation, 129-131
　intersubject variability, 136-137
　theoretical considerations, 122-126
　time dependence, 136-137
Informed consent, 5, 10, 12-13
Institution review board, 9
Irritant testing, 19-40
　animal species, 22-25
　chamber scarification, 34
　dosage, 27
　guidelines for chemicals, 35-40
　ID 50, 34
　IT 50, 34
　mouse ear, 33
　occlusion, 27-28
Irritation,
　acute, dermal, 36-40

Laser-Doppler flowmeter, 31, 108-115
　blood flow measurement, 108-110
　for irritant evaluation, 31, 114-115

Open epicutaneous test, 51, 53, 66
Optimization test, 51, 52

Index

Patch tests, 26-34, 47-72
  allergens (see Sensitization assays)
  animal application, 26-27, 48-49, 69-70
  exposure times, 29, 63-68
  materials, 26
  repetitive patch testing, 33-34
  scoring, 29-31, 70-71
  vehicles, 28-29, 59-60
Permeation methods, 171-181
  application of compounds, 178-179
  calculation of data, 179-180
  diffusion cells, 172-173
  receptor fluid, 175-178
  skin preparation, 174-175
  skin type, 173-174
  variability, 180-181
Persistent light reactivity, 90
Photoallergy testing,
  animal, defined, 94
Photopulse plethysmography, 110-112
Phototesting, human, 93-103
  photomaximization, 96
  phototoxicity, 97
  predictive testing, 102-103
  radiation source, 100
  subjects, 100-101
  substances, 98-99
  variables, 98
  variations, 101-102
  vehicle, 99-100
Phototoxicity testing,
  animal, defined, 94

Sebum excretion rates, 187
Sensitization assays, animal, 47-72
  choice of test, 62-63

[Sensitization assays, animal]
  epicutaneous, 53-55
  factors affecting induction, 57-62
  genetic factors, 61-62
  guinea pig, 48-49
  induction concentration, 63-68
  interpretation, 70-71
  intradermal methods, 49-52
  mouse methods, 69-70
  vehicle, 59-60
Single injection adjuvant test (SIAT), 51-52, 55-57
Skin penetration (see Permeation methods)

TINA test, 51, 55
Transepidermal water loss, 30-31, 191-212
  air current effects, 202-203
  closed cell method, 194, 198
  defined, 191-193
  electrolyte water analyzer, 199
  environmental considerations, 209
  evaporimeter, 30-31, 199-203, 211-212
  flow-through cell, 198-199
  gravimetric method, 193-194, 197-198
  open cell method, 194-197
  relative humidity effects, 206-207
  study design, 209-211
  subject selection, 207-208
  temperature effects, 203-206
TRUE test, 60
Trypan blue, 30, 32

Ultraviolet spectrum, 95